1986

Longman Guide to World Science and Technology

LONGMAN GUIDE TO WORLD SCIENCE AND TECHNOLOGY

Series editor: Ann Pernet

Science and Technology in the Middle East*
by Ziauddin Sardar

Science and Technology in Latin America*
by Latin American Newsletters Limited

Science and Technology in China*
by Tong B. Tang

Science and Technology in Japan*
by Alun M. Anderson

Science and Technology in Eastern Europe
by Vera Rich

Science and Technology in the USSR
by Vera Rich

Science and Technology in the UK
by Anthony P. Harvey and Ann Pernet

Science and Technology in South-East Asia
by Ziauddin Sardar

Science and Technology in the Indian Subcontinent
by Ziauddin Sardar

Science and Technology in the USA
by Albert H. Teich

Science and Technology in Australasia, Antarctica and the Pacific
by Jarlath Ronayne

*Already published.

Science and Technology in China

Tong B. Tang

Distributed by
GALE RESEARCH COMPANY
Book Tower
Detroit, Michigan 48226

Longman
London

SCIENCE AND TECHNOLOGY IN CHINA

Longman Group Limited,
6th Floor, Westgate House, Harlow, Essex CM20 1NE, UK

Distributed exclusively in the USA and Canada by Gale Research Company,
Book Tower, Detroit, Michigan 48226, USA.

First published 1984

British Library Cataloguing in Publication Data

Tang, Tong B.
  Science and technology in China. – (Longman guide
  to world science and technology; v.3)
  1. Science – China   2. Technology – China
  I. Title
  509'.51   Q127.C5

  ISBN 0 582 90056 5

Typeset on Linotron 202 in 10½ on 11½pt Plantin
by Tradespools Limited, Frome, Somerset
Printed in Great Britain by
Butler and Tanner Ltd, Frome, Somerset

Contents

Preface

With an enduring curiosity in how new societies should and actually work, and as an ethnic Chinese, I am always fascinated by events in China. Chinese science and technology have interested me particularly after my becoming first a student of electronic engineering, then semiconductor physics and now solid-state chemistry. Of course, I command a restricted horizon and, apart from my study visits in 1967, 1968, 1979 and 1980, lack of living or working experience in the country proper. Nevertheless, from 1977 onwards I wrote a series of articles on Chinese science in technical and popular journals, magazines, and newspapers. My hope was that any addition to the awareness in the West, however insignificant in substance, may benefit from – and be advantageous to – China's open-door policy which has been emphasized once more. The same hope lies behind the writing of this book.

The book outlines the Chinese development and current status of applied science and high technology, necessarily with omission of details. Occasionally, I have generalized where it would be instructive to particularize, and I plead guilty to any charge of superficiality. However, my sincere apology is offered if an overall picture that can guide researchers or other concerned personnel in their first encounters with Chinese science does not emerge from the book. The subjects covered encompass those that are being treated as priority fields by the government. Basic disciplines like mathematics and astronomy as well as intermediate technologies such as textile, municipal engineering and food processing are excluded. One chapter needs some explanation. Due to historical causes, as of 1983 the province of Taiwan remains under a different administration from the mainland. As the difference is reflected in the policy, organization and indeed the whole social context of science, a separate chapter is devoted to the situation in Taiwan.

Two chapters have been written by invited contributors, to whom my gratitude is recorded here. It must be stated, however, that none of them saw the writings of the others, nor of mine, before the entire text was forwarded to the publishers. I alone am responsible for the structure and the consistency of the whole book. Some of the information in Chapter 1.6 was derived from 'Libraries and information services in China' (*Journal of Information Science* 6 [1983] 21–31). Sources of data or texts in Appendices 1–7 have been individually acknowledged in the context. A note should also be made regarding the Directory of Major Establishments. The information included there is in general adequate for postal correspondence. Thus, for instance, what follows is a correct address:

The Responsible Person
Foreign Affairs Office,
Institute of Physics,
Chinese Academy of Sciences,
Beijing,
People's Republic of China.

I should like to thank Sir Michael Stoker for writing the Foreword for this book.

Tong B. Tang (Deng Tang-bo)

Cambridge

Foreword

Since the door began to open in 1978, many scientists in western countries have established friendly and profitable working relationships with Chinese students in their own laboratories, and a considerable number have themselves visited China to make direct contact with their opposite numbers in universities and research institutes, and some have even allowed a little time for a stroll on the Wall. My own first visit to China was in 1978 when, as Foreign Secretary of The Royal Society, I hoped to negotiate a modest new exchange agreement for scientists, but only after the usual courtesies and protracted bargaining. Instead, Vice Premier Fang Yi, Chairman of the State Commission on Science and Technology, issued an immediate challenge. If, he said, the USA could take a thousand Chinese students now, what was the maximum number he could immediately send to the UK, and how quickly could this be expanded? The door was indeed open.

In the subsequent years of contact with Chinese scientists, most of us have been fascinated, but also bewildered, by the extraordinary mixture of pragmatism, energy, friendship, ruthlessness, humour, inflexibility, backwardness, achievement, and once again pragmatism. A systematic account of the organization of science and technology and its background in present-day China was badly needed, and this is why Dr T.B. Tang's contribution 'Science and Technology in China' is such a welcome and valuable addition to the Longman Guide to World Science and Technology. Not only does it provide a factual account with useful statistical information (teacher/student ratio 1:4 indeed!) but it also helps us to understand the social and political basis for the inconsistencies, which are unlikely to disappear in the coming years, despite the tireless probing and experiment in the effort to regain China's historical pre-eminence in science.

The book will be particularly helpful to scientists who are supervising the work of Chinese students in this country: for example, it provides homely advice on what to do when things go wrong. And future visitors to China who take Dr Tang's volume with them will be far better equipped than I was.

Sir Michael Stoker CBE FRS FRSE FRCP

1   Introduction

China has a long prehistory; the remains of Homo erectus have been discovered at Yuanmou in Yunnan province and dated with palaeomagnetic methods to be 1.7 million years old. The span of recorded history covers an uninterrupted interval of thirty-five centuries. For the scope of this book, however, the past three and a half decades constitute the period of attention. Since 1949 power has resided within the Communist Party. Science and technology have developed within a socialist framework, which is as significant a determinant to their contemporary characters as the totality of Chinese traditions. In this chapter we examine their economic context (capital), planning (time), organization (system), social setting (people) and international aspects (nations).

1.1   Demographic and Economic Background

China has an area of nearly 9.6 million square kilometres and a coastline extending to 18 000 kilometres. Included in the territory are more than 5 000 islands, the biggest two of which are Taiwan and Hainan. Away from the eastern seaboard the country is mostly mountainous: a third of her land area is at altitudes exceeding 1 000 metres and a fifth, over 5 000 metres. Three major waterways flow from the west to the east: the Yellow, the Yangtze and the Pearl River systems. In latitude China spans from 4 to 54 degrees north and, in the main, enjoys a sub-tropical to temperate climate, with an annual frost-free period nowhere less than 120 days. Further geographical data may be found in Table 1 of Appendix 1 at the end of this book.

An official census conducted in 1982 put the population at 1 014 millions. The mean rate of increase over the past decade is estimated to be a modest 1.2 per cent per annum, and partly attributable to the continual improvement of life expectancy which, according to United Nations statistics, now reaches 70 years. Nevertheless, the government has imposed ruthless measures of birth *prevention* (see p 238). Six per cent of the Chinese people belong to one of fifty-five ethnic minorities, the rest being of the Han race. The breakdown of the population by administrative divisions is shown in Table 2 of Appendix 1. Xizang (Tibet), Xinjiang and Inner Mongolia autonomous regions and Qinghai province are huge but sparingly inhabited. Disregarding them, Sichuan province is the largest in both area and population. The three most populous cities of Beijing, the national capital, Shanghai and Tianjin together with respective surrounding counties have been administratively made into municipalities, which command provincial status. A municipality is subdivided into urban districts and rural counties. Below the district, local, voluntary administrative units exist on the 'neighbourhood' and then 'street' levels. A region or province consists of prefectures, themselves made up of counties and cities. Several communes (towns) comprise a provincial county, and are in turn composed of villages (formerly production brigades).

About 85 per cent of the Chinese labour forced is engaged in agriculture and fishery, which accounts for a sixth of the Gross National Product. Agriculture is important also in that it provides 70 per cent of the raw materials required by light industry, the latter in turn accumulating over half of the capital needed for the expansion of heavy industry. On the whole the growth rate of the Gross National Product has been high, the annual figure being 7.2 per cent averaged over the thirty years between 1950–81.

The development of China's planned economy has broadly followed a series of Five-Year Plans (FYP), the first of which appeared in 1953. It was fulfilled ahead of schedules and a new one commenced in 1956. This Second FYP met with mixed results due to unforseen disruptive factors such as problems with intellectuals, and an abrupt development in the form of the Great Leap Forward, a mass mobilization campaign that foundered owing to bureaucratic abuses and natural calamities like floods. The difficulties were compounded by the actions of Nikita Khrushchev, who cancelled all the 257 Sino–Russian scientific cooperation programmes, tore up the 343 mine and plant construction contracts, and recalled the 1 390 technical advisors. It is apparent that both the First and the Second Plans channelled more investments into industry than into agriculture, as may be verified from Appendices 2 and 3. This preference followed the pattern in the USSR. Generally speaking, Soviet influence was strongest in those realms where the Party lacked experience, ie state-to-state diplomacy, central government organization, the institutionalization of technocrats, higher education and scientific research, as well as the management of the national economy.

The end of the 1950s marks a watershed in Chinese economic policy. By this time socialist collectivization, acted on by the peasants' own initiatives and (in this case) not by pressure from above, was basically complete in the countryside, so that political reorganizations were not expected to raise agricultural productivity as much as before. Moreover, the faults of the Russian model, in which heavy industry takes precedence over other sectors, came into sharp relief in the picture of general economic failures then prevailing. Not only were there famines, industrial expansion ground to a halt because of the poor performance in agriculture, which provided food, raw materials and surplus that could be turned into investment capital. It was in 1961 that the directive 'agriculture is the foundation of the economy' was issued. The principle of putting agriculture first has been upheld in all succeeding FYPs. In the slogan of Four Modernizations, enunciated in 1975 but now more than ever valid, the order of priority is agriculture, industry, defence, and science and technology.

INTRODUCTION 3

The period from 1961 to 1965 was devoted to readjustment and no FYP was in effect. The Third FYP which should have lasted from 1966 to 1970 existed merely in name due to the intervention of the First Great Proletarian Cultural Revolution, commonly called the Cultural Revolution. The Fourth and the Fifth, in force during the decade 1971–80, are outlined and compared in Appendix 4. The country is currently in the midst of the sixth FYP, sketched in Appendix 5. Both this and the next plan (see Appendix 6) are formulated with reforms, restructuring and consolidation in mind. Apart from agricultural advancement, infrastructural developments in energy and transport also receive emphasis; the 1990s are predicted to witness a faster rate of growth, the aim being to quadruple the annual gross national output value in 1980 by the year 2000. It was announced at the end of 1983 that the 1985 production targets (as outlined in Appendix 5) were reached two years ahead.

The character of China's economy underwent drastic changes at the end of the 1970s, when many of the radical policies brought about by the Cultural Revolution were reversed. Materialistic solutions are now sought for economic growth, which is viewed largely as an end in itself. The last year or so saw a shift towards the centralization of power into the hands of the commissions and ministries, at the expense of provincial and local administrators though, apparently, not of the managers of individual enterprises. Market regulation through pricing and other voluntary mechanisms is now tolerated, but public ownership and planned economy remain mandatory.

1.2 Science Policy

Progress in science and technology has kept in step with economic policies, and has also been highlighted by a progression of state plans. The first, a twelve-year plan (1956–67), was started at the same time as the Second FYP; many of China's older research institutes were established in that year. Programmes on atomic energy, jet and rocket engines, electronics, computers and other areas were included. The plan saw completion five years ahead of schedule. Thus, in 1963 a new ten-year science plan was drafted, but its implementation was greatly affected by the outbreak of the Cultural Revolution. At the National Science Conference convened in March 1978, the government announced an eight-year project for the development of science and technology (see Appendix 7). More recently, a plan to cover the period ending 2000 has been mentioned, but no details are known.

In general terms, however, this latest fourth science plan is portrayed as reflecting priority decisions that are based on economic needs and designed to encourage the practical application of research results. Such goals, which have marked Chinese science policy ever since the early 1960s, are currently reiterated in almost all relevant speeches and public statements made by state leaders including the party chairman or vice chairman and the premier. This concern at the highest government level is not as unusual as it would be in a Western context. In China the politicization of all spheres of social activities is explicitly acknowledged, so that political directives can and do guide their directions of change. On the other hand, the continual stress on the economic exploitation of scientific research demonstrates the lack of change in the situation. Indeed, the major problems that face the country's science policy makers may be examined through a discussion of the 'four transfers' being urged in the Chinese media. This refers to the transfers of science and technology 1) from laboratory to production, 2) from the military to the civilian sector, 3) from urban centres to the rural backland, and 4) from overseas to China.

Let us discuss the first. In ancient times the Chinese appeared to be weak in moving from invention to innovation. They invented things such as gunpowder, gear wheels, cranks, piston-rods and blast-furnaces, none of which came into widespread use. There were even many examples where discoveries were allowed to die out, ie in seismology, horology and iatro-chemistry. This phenomenon may be rooted in the traditional attitude of revenerating 'pure' knowledge. In China today, applied research is reckoned to absorb 90 per cent of civilian science funds, yet a mere 10 per cent of the research results can be utilized promptly in production, as opposed to 80 per cent in the USA. The whole situation looks comparable to that in the Soviet Union, which is similarly plagued by an inability to get the greatest return from its domestic industrial research efforts, partly due to an engraved bias in favour of theoretical work that can be traced to before the October Revolution. However, the Chinese for a long time, and especially since the Cultural Revolution, have instituted unusual arrangements to bridge the gap from laboratory to production, such as the incorporation of pilot factories within science centres and university departments. Unfortunately, two chief factors weaken the link.

One factor, dubbed the 'swarming of bees' by the Chinese themselves, refers to the tendency among some scientists or institutions to concentrate on topics that are fashionable, causing serious repetition and waste of effort. As an example, no less than 980 groups in the country are studying haplid seed breeding. Such bandwagon tendences also exist in the West, of course, but they are counteracted by the threat of grant termination unless justifying results are exhibited; in China researchers do not (yet) have to 'publish or perish'. The duplication of effort leads to a great reduction in the number of projects that yield independent applications.

Furthermore, the fashion in research may be insulated from economic realities. For instance, the ambitious programme on particle accelerator announced in the 1978–85 Science Plan may reflect the influence of high-energy physicists both at home and from abroad, rather than the identification of long-term applications. It is said, however, that the objectives in the 1986–2000 plan adhere firmly to the needs of national construction. Other administrative measures to help production include the adoption of a 'responsibility system' whereby contracts can be signed between a factory and a research team, but only for work proven by the latter to have specific relevance. Some factories themselves host 'technology development centres' with temporary staff from outside. Beginning March 1983, the Chinese Scientific and Technical Association set up a subsidiary called Science and Technology Consultancy Service, to function as the umbrella organization for over 500 existing user-orientated research units. Earlier, in January 1982, the Academy of Sciences established a science fund which supports projects with well-defined aims and periods of completion (normally two or three, and at most five years). The fund is unique in being distributed directly to individual researchers rather than through their parent institutes, and overlapping projects will be evaluated against one another. As a parallel measure, coincidentally, local branches of the People's Bank of China are offering loans to finance the expansions of manufacturing facilities that put the fruit of research into practice. In the past, exclusive preoccupation with the fulfillment of production quota often deterred the introduction of novel technologies as this involved an extra outlay of capital and disruptions to ongoing production processes.

The second transfer refers to the military-civilian coupling, or the lack of it. From the mid-1950s onwards, China gathered a huge army of scientists working under the State Commission of Science and Technology for National Defence. Its personnel occupy a privileged position, like being placed in a sanctuary free from interference by the Cultural Revolution. Their overall performance has been brilliant: the missions to make the first atomic and hydrogen bombs, long-range missiles, and artificial terrestrial satellites were all accomplished in remarkably short times. Regrettably, the expertise they acquire rarely diffuses to industries outside military control. There are two reasons for this. One is the obsession with secrecy surrounding classified work. The other is bureaucratic compartmentalization, which hinders any horizontal flow of information. This latter aspect describes Chinese institutions generally. Government ministries sometimes behave like 'independent kingdoms'; two groups segregated organizationally, but researching on the same topic and perhaps in neighbouring buildings, seldom collaborate and sometimes guard against each other. Cooperation is more tenuous still among segregated groups that work in different areas which would benefit from cross-fertilization. Technical exchange has gained momentum, at last, as academic periodicals, many of which were suspended during the Cultural Revolution, resume publication, while at the same time more titles have appeared. Conferences now meet frequently after local learned societies under the Chinese Scientific and Technical Association resumed activities. The newly created Science and Technology Leading Group (see following section) may forge a stronger link between the military and the civilian sectors. A push from a high political echelon is needed, but may be insufficient to ensure interdepartmental cooperation.

The third of the 'four transfers' concerns that from cities to the rest of the country. China is vast and has been industrialized for little more than a century. The majority of her population live in villages and small towns, the economic potential of which await full realization. To raise the scientific level of peasants and town dwellers is vital to the diversification and development of the rural economy. Unfortunately, albeit inevitably, few advanced research centres are found outside Beijing, Shanghai and the provincial capitals. This means not only that the rest of the country cannot gain easy access to knowledge of science and technology, but also that

MAP OF THE PEOP

PUBLIC OF CHINA

problems which arise there may easily be overlooked. Fortunately resources are now being diverted to improve the situation. A national conference met in October 1982 to discuss ways of enhancing scientific work in towns, and three months later another was convened on scientific work in the countryside. The enduring campaign to popularize science and technology has been stepped up. The local branches of the Chinese Scientific and Technical Association are in formal charge of science popularization, of which more in Section 5.

The last transfer relates to the import of expertise from foreign countries. Consideration on this point will be relegated to Section 6.

1.3   Science Organization

In the previous section the Chinese Scientific and Technical Association has been mentioned several times. Inaugurated in 1958, the association kept a low profile during the Cultural Revolution but was revived in 1978. As mentioned already, it is the national coordinator of consulting services, the 'home' of all Chinese scientists through their membership of specialist societies, and the promoter of science popularization. It is not, however, a source of research grants.

The central government organ entrusted with the distribution of funds is the State Economic Commission, which prepares annual estimates for all ministerial spendings. The funding pattern is fixed on the recommendations from the State Planning Commission. The formulation and implementation of science plans lie in the hands of the State Scientific and Technological Commission, which started its existence as part of the Planning Commission in 1956, at the beginning of the twelve-year science plan. Two years later it became independent. Its power has seen ups and downs over the years but, at present, it is certainly the most comprehensive organization devoted to the management of civilian science, and functionally is the top of the hierarchy of research institutes and laboratories, although administratively it interfaces with them only indirectly via ministries. Its budget adds up to approximately half of China's allocation for science and technology, the rest of which goes to provincial scientific and technological commissions and to the State Commission of Science and Technology for National Defence. As in other countries, extra expenditures on military research and development are probably hidden under other names.

Below the commissions are the ministries and special agencies. The Ministry of Education oversees all schools and colleges (next section). It may be joined by another ministry in exercising dual control over particular specialized colleges. Some ministries have academies to which research institutes are affiliated, as will be individually described in appropriate chapters of this book. These academies are small when compared to the Academy of Sciences, but work overlap is unavoidable, although investigations related to technology in a strict sense are not typically conducted under the auspices of the academy. Institutes are known to have been transferred out of the academy and then back again as its fortune fell and rose. A few years ago there was talk of splitting it into two, one for fundamental natural sciences and the other for applied, but it has obviously succeeded in opposing the idea.

The academy was founded as a subordinate to the Ministry of Education on 1 November 1949 but, in 1954 (concurrent, perhaps significantly, with the commencement of the First FYP), it was elevated to ministerial status. By 1965 the number of institutes under its aegis grew to 106. The Cultural Revolution brought about a dramatic contraction as a result of decentralization and the denigration of basic research. In 1973 only 53 institutes remained, the rest having been amalgamated, closed down or transferred to another academy. Growth has recurred since 1977 and it now operates some 120 institutes. They are grouped under Mathematics and Physics, Chemistry, Biological Sciences, Earth Sciences, and Technological Sciences. In cities where they concentrate, twelve Branch Academies have been established: in Beijing, Shanghai, Xi'an or Xian, Hefei, Chengdu, Nanjing, Kuming, Guangzhou, Changchun, Shengyang, Wuhan and Lanzhou. Apart from these, eight instrument factories, the University of Science and Technology of China located in Hefei and its graduate school in Beijing, are also under the jurisdication of the academy. So is the Science Press. This publishing house puts out books of all levels plus around a hundred primary and review journals, including 'Scientia Sinica', which is perhaps the most prestigeous multidisciplinary periodical in China.

Active research is not confined to universities and institutes. Laboratories attached to some state enterprises, as well as establishments supported by offices of science and technology under provincial authorities, carry out work relevant to local needs. These units are numerous and do not, as a rule, figure in central government budget. 'Mass' scientific stations also exist in the form of agronomic, meterological and seismological networks in rural regions. They play an indispensable role in data collection and, to a smaller extent, data reduction.

A Technical Economic Research Centre was set up in 1981 in Beijing under the State Council, which is the highest decision-making body in central government. The centre acts as a consultancy, with professional personnel drawn from the permanent staff of other research institutes, whose work for the centre is on a part-time basis. In 1983 a Science and Technology Leading Group was created within the council, with the following *ex officio* composition. Head, the premier; deputies, the minister in charge of the Scientific and Technological Commission and a vice minister-in-charge from the Planning Commission; others, the president of the Academy of Sciences and vice ministers of Education and of Labour and Personnel. (The Bureau of Scientific and Technical Cadres in the Ministry of Labour and Personnel supervises the assignment of college graduates to posts in research and educational establishments. See the next section). The express tasks of the Leading Group are to map out long-term science development, to define policy on technology imports, and to function as the central director of military and civilian research. With membership drawn from such a high level, the group has to restrict its work in breadth and volume, but its formation underscores the importance attached by the political leadership to science.

1.4   Tertiary Education

Over the past four years China's spending on education rose by ten per cent per annum on average, or about half of the corresponding rate for science. The ten per cent rise represents a very modest per capita increase, in view of the burgeoning pupil and student populations. Almost a million and a half now attend regular universities compared to the total number of graduates during the past three decades which were less than three times that amount. In 1980 and 1981 280 000 new students were enrolled; the intake is projected to be 400 000 in 1985, but even then will still be far from adequate in relation to national need and national population. Part of the expansion has been made possible by the admission of commuting students: before 1980 all students lived on campus. The dormitories were free but rent is now charged, though they remain crowded (cf p 37). It was small consolation that two years ago a soft loan of US$233 million was negotiated with the World Bank, for the purpose of ungrading the facilities in institutions of higher education. Beijing and Qinghua, the top two universities, got $20 million each. It is perhaps fair to say that, while the importance of science as a productive force is well recognized in China, the role of education in the progress of science is insufficiently stressed. The universities should serve as centres of both teaching and research but, in practice, are much poorer than the institutes under the Academy of Sciences. Indeed, most overseas Chinese scientists returning to work on a long-term basis have ended up in academy institutes, and few in universities.

Over 700 institutions cater for tertiary education in the country. Among them there are, in round figures, 50 comprehensive universities, 250 polytechnics and specialized engineering colleges, 90 medical institutions, 80 agricultural schools, 120 teacher training colleges, 15 each of foreign language and physical education schools, and 40 each of establishments for art and for economics, trade, law or diplomacy. The total number of teachers just exceeds 250 000. The courses offered by medical colleges last six years, in polytechnical universities, five; in agricultural schools, four. They also last four years in comprehensive universities which incorporate science instruction. Engineering and science are taught in separate institutions. This system has disadvantages but is now too well established to be modified easily. A major exception is the University of Science and Technology of China. It has both science and engineering departments, and belongs to the Academy of Sciences. The Ministry of Education oversees all the other institutions, but for medical, agricultural and specialized engineering colleges control is exerted jointly with the ministry in charge of a particular field.

Some eighty universities and colleges are designated 'key' institutions, which enjoy priorities in the allocation of funding, staff and students. Among them, twelve form a league of the '*creme de la creme*'. They are all universities of science or technology and, grouped under their cities of location, are:

(Beijing) Beijing, Qinghua, Beijing Normal;
(Shanghai) Fudan, Shanghai Polytechnical;

(Xian) Xian Polytechnical;
(Nanjing) Nanjing;
(Hefei) Science and Technology of China;
(Wuhan) Wuhan, Central China Polytechnical; and
(Guangzhou) Zhongshan, South China Polytechnical.

An institution of higher education is organized into a number of departments, each containing several 'teaching and research groups'. These groups prepared curricula, time tables, textbooks, stencilled materials and examination papers, and individual members carry out research with the help of graduate students and teaching assistants. The faculty staff consist of lecturers, associate professors and professors. Matching in grade the assistant, associate and full researchers of the Academy of Sciences, they total respectively 4 300, 20 000 and 118 000 in number. As a rule the professors of a university form an 'academic committee', the legislative body for all matters outside the terms of reference of the resident Party office. Graduates obtain degrees if they pass with distinction. An Academic Degree Committee, created in 1980 by a State Council decree, has delegated its power to ten Appraisal Boards (one for each different discipline) to lay down regulations for the award system. The first doctoral degrees were granted in the summer of 1982 amid pomp and circumstance. The regulations stipulate that persons engaged in self-study can take examinations for a diploma, which is equivalent to a degree (see below).

The initiation of the degree system is one of the many reforms that have swept across the education scene; before and during the Cultural Revolution no degrees were conferred. Indeed, education in China is a sensitive barometer of the political climate – more sensitive even than science policy. Thus, for the past three decades, the proportion of science students in higher education has hardly strayed from seven per cent and engineering graduates, form one-third of the total, but their training has undergone drastic changes. Unlike the West, China needs to infuse new morality into her society, if her present political system is to become a firm social foundation based primarily on education. During the Cultural Revolution the novel social values were inculcated by the shortening of courses (from four to three years in universities). Labour was used as an educative agent and students were required to participate in industrial or agricultural production. Spare-time colleges attached to factories and communes mushroomed, so that the distances between the classroom and the society and between manual and intellectual work were narrowed. Politics took command to the detriment of academic standards. Streaming disappeared, no key school was recognized, university admission came to be on the recommendation of 'the masses' but not on the basis of examination results, and worker/peasant representatives were posted to share power with professionals.

Most of these measures have now been reversed. The sole exception is that informal classes run by factories and communes are still encouraged. Technical personnel are in short supply in China, amounting to a mere six million out of a population of one thousand million. (If research scientists are counted, then the figure is short of half a million.) As an alternative to formal tertiary education, spare-time education for people from factories or communes is an effective means of boosting the quantity of professional personnel. The enrolment in adult higher education now already exceeds that in regular full-time colleges and, by 1990, may reach three million. In this context the vocational and technical schools should also be mentioned, although they relate to secondary rather than tertiary education. Their students number 2.5 million, or 23 per cent of the aggregate figure for middle schools, and are intended to take up the role of technicians and skilled workers. They were, surprisingly, abolished during the Cultural Revolution, when senior secondary schools went through a seventeen-fold lopsided expansion.

There is room for more reforms for China's universities. Working hard does not suffice if one has to work against the system. One characteristic is rigidity in their running. At present students once admitted cannot easily change departments, receive trainings which are highly specialized, even by English standard, and are allowed little freedom in selecting jobs after graduation unless they boast of personal connections. (See following section regarding job allocation.) Besides early specialization, a product of Russian influence, there is uniformity. Curricula and textbooks rather lack diversity although, after recent reforms, most faculties now offer many 'unlimited elective' courses which any student is free to choose; other courses are either compulsory, or 'limited elective' (meaning that students have to take a specified number of courses on the subject they are reading). Still, the emphasis is on teaching not learning. As specified by the Ministry of Education, a science student has to be provided with at least 30 hours of lectures per academic week. The tutorial style of 'lecturing' is reflected by the superficially excessive ratio of staff to students, which is typically 1:4. Other consequences are that students may lack initiative or at least be inarticulate, and that teachers may have to neglect research.

Scientists in China have enjoyed a spectacular rise in esteem, though not what some regard as a speedy increase in salary. Their social status can be judged from the media – it has become fashionable to talk or write about their life, work, or even their lovelife. As could be expected, aspiring youths now regard reading a science subject in a key university as the best bet – and theoretical studies are often more sought after than experimental ones. In contrast, during the 1970s, a popular option was a career in the army.

College entrants are now picked on a much narrower set of criteria than before. It depends mainly on performance in the qualifying examination held nationwide in every summer, although health and character are taken into consideration. Newspapers have reported on the admission of a student with criminal records, implying the possibility but improbability of such cases. The competition is tough. There is just one place for every 20 senior middle school leavers. Even in Beijing, where schools are better, on average only one in ten gets in. Worse still, the competition may not be fair, for some are favoured by circumstances. There are those who go from one key school to another, all the way from kindergarten to university. Out of the 66 million school pupils in China one million are in key schools. Of this million, 15 per cent are from families of industrial workers, a percentage which roughly equates with that of workers in the national population, but only 5 per cent are from peasant homes. The rest are children of intellects and cadres who themselves constitute less than 5 per cent of the population. Exactly the same social composition is found among tertiary students. Even within ordinary primary and secondary schools, there was streaming of brighter pupils into special 'fast classes', although by and large this practice has disappeared over the last year or so.

The Chinese are taking action to combat the inequality in educational opportunity arising from wider social forces than just personal capability. The marking of entrance examination papers is slightly adjusted to positive discrimination against candidates from schools in rural area. Moreover, the Ministry of Education has announced a scheme by which the self-taught can stand on an equal footing with the formally educated. Anyone, regardless of age or previous schooling will be able to obtain a diploma recognized as equivalent to a first degree after an examination. These 'open' examinations are to be attempted at one sitting or over a period on a credit system. Furthermore, the press is full of stories about the promotion of people who have shown abilities but possess no formal qualifications. All these measures constitute a step towards the abolition of 'the bureaucratic baptism of knowledge', or examination, which can so easily degenerate from an assessment of understanding into a test of recall.

Research students in China now approach 15 000 in number. They can be placed under three categories: the tiny group of those who started their undergraduate courses before 1965, the big but shrinking group of those who started during the Cultural Revolution, and the small but enlarging group of those who started after 1978 when entrance examinations were reintroduced. During 1967–77 there were no examinations and no post-graduates. Students in the second group are unique in that they had been sent in from factories and communes under a quota system. Called 'worker, peasant and soldier' students because of their past experience irrespective of family background, they had not been selected on the exclusive criterion of (or occasionally with any regard to) academic standard. Indeed, examinations are being offered to them in individual cities, and should one still fail after three attempts he or she is disqualified from their current post. However, on the more positive side it goes to show that someone who has once been made a worker or peasant can turn into a scientist. This assurance is the real basis for the great leap forward of Chinese science.

All in all, China's educational policy has taken many sharp turns. It now appears to have entered an extended period of evolution rather than revolution, though people thought the same just prior to 1966 and were proved wrong. The overriding force behind the changes is not technocratic but political in nature, and stability will be guaranteed by the continuity of the Party leadership.

1.5 Human Resources and Science Popularization

A country's progress in science and technology is chiefly determined by the level of her technical manpower, which in turn is ultimately dependent on how much of the human potential can be realized under its social system. China has a unique handicap in this respect. In 1949, when the Communist Party gained power, there were about two million intellectuals. Most of them originated from or had been in the service of the overthrown privileged classes, and had benefited from education in the big cities (capitalist strongholds) on the eastern seaboard, in Japan or in the West. Their enthusiasm for socialist China was maintained by patriotism until confrontation with the succession of political campaigns from 1952 onwards, whence those who could not remould themselves in line with the new law and order became passive in these movements. Their passivity turned into antagonism in the Anti-Rightist Campaign of 1956 and in the Cultural Revolution.

The drive, in effect if not in motivation, during the Cultural Revolution was to dilute the professionalism of scientists. Education then aimed, as we have seen, less at academic excellence than at social homogeneity. In research, the experiments of 'going out' and 'inviting in' were initiated. 'Going out' involved scientists being 'exiled' by rotation to factories or communes and participating in the solution of difficulties encountered in production. 'Inviting in' meant opening research facilities to workers or peasants who came with practical problems – in short, the converse of going out. The delineations of organizations and positions with their attendant status differentials and divisions of labour became blurred.

After the Cultural Revolution the status, job titles and responsibilities of scientists were quickly restored, and their working conditions gradually improved. However, there are still factors that hamper the harnessing of their full abilities. One of the most significant is connected to job allocation. Normally they are assigned to their posts by the Scientific and Technical Cadre Bureau, and once done, any attempt to transfer to other duties will be met with insurmountable obstacles. It is considered that this 'unified distribution' system is essential for the operation of a planned economy, as it provides a stable and predictable labour market. However, the allocation may not be entirely rational, and many are assigned unsuitable jobs. Indeed, some have aptly attributed the failure to employ university graduates to be one of the immediate reasons behind the success in the launching of the Cultural Revolution. Fortunately, new arrangements are being experimented with that will help remedy the situation. A few institutes have started advertising positions, to which scientists working elsewhere may apply on their own initiative. Direct contacts between employment units and educational institutions are encouraged. The former specify the quantity and qualifications of the graduates required, and the latter will suggest assignments within the constraints already imposed by the demand of the Cadre Bureau. Alternatively, a personnel training contract can be drawn, stipulating the amount of college fees to be borne by the prospective employer. People attached to research centres are now allowed to take up outside temporary jobs, usually for a period between six and twelve months.

Certain reforms are, paradoxically, either resisted or quickly implemented because of their resemblance to Western practice. China has had little experience of capitalism. Hence the possibility, for instance, that democracy is confused with individual liberalism by some intellectuals as well as by some bureaucrats. This partly explains the difficulties encountered when one tries to change jobs, or when two people fall in love while studying, but upon graduation are not allocated work in the same town, resulting in 'tales of two cities'. The same confusion also explains why sometimes democracy is sought after, or protected, in the wrong way.

On the other hand, policies of personal (rather than collective) material incentive have been implemented quickly, perhaps because they belong to some image of the successful West. Authors are paid for journal publications, and monetary prizes are given to professors deemed to have made important contributions. The Academy of Sciences has gone to the extreme in putting into effect, on a trial basis, the arrangement whereby personnel who do not complete their work assignments on time have their wages reduced or cut off! The immediate reason for research centres to accept contract work is to secure extra incomes which can be passed on as bonuses to the staff. Undoubtedly, scientists are not well off, often getting less than workers of corresponding work ages. They have to worry about the novel (for China) phenomenon of inflation and at the same time fight against the new trend of being expectant consumers. So although many are motivated by socialist ideals or by the reward of science *per se*, low pay has become a subject for loud complaint. This dissatisfaction is one of the main reasons for the ongoing brain drain. Each year nearly 55 000 legal 'emigrants' went to or passed through Xianggong (Hong Kong), a large fraction of whom were scientists, engineers and doctors. Most of these professionals are now discovering that being underpaid is less depressing than being unemployed or underemployed.

So there are many paradoxes. The country is acutely short of engineers and scientists but China still does not make the most of her university graduates. Young people compete fiercely to study science while scientists complain of low pay. Yet, contradictions can be made beneficial for development. The Chinese know well to avoid becoming socialist in form, capitalist in aspiration and feudalist in substance.

One armoury in the battle to wipe out feudalism is the grasp of scientific knowledge in the hands of the broad masses. In ancient times non-hereditary mandarins were selected by imperial examinations which theoretically anyone could take, so that traditionally the whole populace has been conditioned to seek the advantages education offered. This deep-rooted exaltation of learning as a road to position makes it necessary today to watch out for professionalism and authoritarianism, against which the popularization of science will help fight. In the last 35 years, insistence on a combination of professional and populist approaches has led to a people's health care system instead of a city-orientated medical service. What is attained encompasses not only 'science for the people', but also 'science by the people', as embodied by the barefoot doctors. Similar achievements are made in such diverse fields as seed selection and plant breeding, meteorology, earthquake forecasting, as well as in archaeology. As a result of the unique policy of 'walking on two legs', namely the adoption of traditional techniques and modern methods, rural industrialization has made a start.

Small-scale factories have been set up by villages in the spirit of self-reliance, ingesting little or no government investments, to satisfy the local needs for farming implements, chemicals, etc (by making the 'three machines and one pump', 'four acids and three alkalines', etc).

The means of popularization include radio broadcast and films, lecturing by mobile 'bare-foot' science teams, posters, magazines and books. An Association of Popularization Writers has existed since 1980. In China science fiction, of which some 40 000 titles have been published since 1976, is meant to be as instructive as non-fiction but escapism is critisized. While at present agricultural and industrial exhibition halls, aquariums, planet-ariums and museums are confined to big cities, as are educational displays on street-walls and in shopwindows, similar though more modest facilities are being built or designed in a number of smaller towns. On the commune level, 'scientific and technical service centres' have sprung up which help to disseminate technical information. The national microwave network is expanding gradually and a communication geostationary satellite may soon be orbited; their functions will include the opportunity for people in remote areas to enrol in 'broadcast' universities. At present less than a third of the rural population can receive television via conventional ground networks.

The end of popularization is the elimination of outdated ideas in the social consciousness of the whole people. It counters social polarizations by narrowing the differences between city and countryside and by reducing the alienation due to social divisions of labour. In the final analysis, it is a mass movement to emancipate the minds of workers and peasants by encouraging their inventiveness in the struggle for production. There have been, both in China and in the West, many cases of important contributions by amateur scientists who have had little formal education. Its outcome will probably depend not so much on funding and other material conditions as on popular enthusiasm. Apparently, in great contrast to most people in the West, the rank and file in China do look up to science as a social activity capable of improving their quality of life, and at the same time look down on science as an alien force because never again will they feel powerless in shaping their own future. Indeed, one should argue for greater popularization, and the demistification of science in the West, if only to revive the popularity of science subjects in schools, to stem the flood of occult 'sciences', and to impress upon scientists their social responsiblity.

The move towards popularization will affect the Chinese direction of scientific progress. Science possesses a historical dimension. It reflects not only the structuring of the objective world, but also the functioning of the subjective mind and hence existing social relations. In the West, its socialization has taken place since the turn of the century and reached fresh heights after the last World War. In China, from the beginning, the management of science has been regarded as a political process. There, as the scientific level of the entire population is raised, democracy in science can be contemplated. Then, techno-economic issues like that of nuclear energy will have the chance of being democratically debated.

1.6 International Exchanges and Information Management

Chinese colleagues are more likely than not diligent, purposeful, cautious, modest, and friendly or at least obliging. This much, it is safe to assume, is a feeling common to all in the West who have worked with researchers or students from or in China. Regarding wider issues, however, it must not be forgotten that most foreigners come into contact with, and learn the opinions of, only a small cross-section of the Chinese social spectrum. The Chinese impressions of Westerners visiting their country are more diverse if not more transient. Scores of scientists from the United Kingdom have made short- and medium-term visits under an agreement between the Royal Society and the Academy of Sciences. By a separate protocol signed by the Ministry of Education and the British Council, students are also exchanged.

In the 1950s, while under the blockade orchestrated by the USA, China conducted scientific and technical exchanges and cooperation primarily with the USSR and the Eastern Bloc. During the Cultural Revolution, similar relationships were established with East European, Asian, African and Latin American countries, and from the mid-1970s onwards, with Japan and Western Europe, but only slowly at first. Since 1978 agreements on exchanges and cooperation have been sealed with the governments of France, Italy, West Germany, Britain, Sweden, the United States, Finland, Denmark, Belgium, Luxembourg, Greece, Australia, Norway and Japan. As direct scientific interaction between China and Western nations has intensified only recently, all parties concerned are eager to review their experience. An unofficial committee of representatives from European, North American and Japanese scientific organizations has been formed to discuss relations with China.

The Chinese attitude may be viewed in the broader context of their modernization programme. Although China, mindful of the experience with Russia, is unlikely ever again to mortgage her independence, she is now predisposed to using foreign expertise – and capitals – as valuable, albeit subsidiary, aids. One way has been to do it at the lowest level, by buying finished products for immediate distribution, or, at an intermediate level (and perceived as the 'Japanese method'), by importing whole industrial plants or complete blueprints of components. However, she has encountered difficulties in this approach, including problems with product compatibility, the inadequacy of absorptive capacity, alien production relationship, and ill-fitting infrastructure like power supply or transport service. These difficulties are accentuated by bad inter-ministerial coordination and crippling bureaucracy. In a few cases she was inveigled into building plants on unsuitable sites or acquiring inappropriate technologies. As an illustration, four large-scale integrated circuit production lines were separately purchased by different officials in Beijing, but for a long time none was made to work, while Chinese scientists and engineers mumbled that given research and development fundings they could make the equipment themselves.

The belief is now as firm as before (as after the break with the Soviet Union in 1960) that a better policy, even in the short run, is to find methods of upgrading selected existing plants and then apply the tested methods to other similar installations, and that new production processes should be developed indigenously rather than imported whenever possible. China relies on her own scientists to generate new knowledge upon which innovations and discoveries are based. Such an effort will benefit from cooperation at the highest level, namely scientific exchanges with the industrially advanced West. Thus while China is importing less high technology, she is sending out more and more students. The number to Britain may, though, remain static at around 300 a year. Proportionally many more are going over to the United States, where they pay less fees and may even obtain scholarships or assistantships. Among the Chinese students who are now abroad, some 80 per cent are in science and technology subjects. Of these about 15 per cent are doctoral students and 15 per cent undergraduates, with the rest doing courses shorter than three years. It is anticipated that research students will not be sent abroad in large consignments once indigenous supply reaches 20 000 year.

In general Chinese overseas students are extremely dedicated to their studies. Their burden of expectation may, however, turn into overt pressure if other people are slow to realize that academic performance is affected by additional factors unrelated to personal ability. A doctoral student may, for example, simply be unlucky in his experiments. The supervision he gets may be less than brilliant. Equipment problems may require solutions which are not only technical but also involve departmental politics. No PhD project, indeed any research plan, is absolutely safe. Whatever the causes, a temporary setback must not result in self condemnation (there was a sad case of suicide in the US). Conditions of work also comprise emotional well-being. It was not until 1981 that the decision was finally taken permitting students to make a trip home to see their families every two years. Some officials may be indifferent. Indeed, back home husband and wife may have been assigned duties in different cities and be able to see each other only once a year. Most of such sacrifices stem not from any absolute contradiction between common and personal interests – the real cost to the central government incurred by the return trips of students is minimal, given good cooperation between the ministries and the national airline. It is rather more possible that they result from the lack of interest on the part of some bureaucrats in personal welfare other than their own. The same may be said of several known cases concerning students studying abroad who, because of personal affairs that appeared perfectly legitimate to an outsider, deserved special freedom of action but whose plights were ignored. But then, an official may be reprimanded for actions outside the rule book, but usually not for inaction!

The above comments should not, of course, obscure the fact that the majority of Chinese studying abroad are doing very well and, more crucially, their ethos are gaining much appreciation from the host peoples. One does not report that an undergraduate obtains a double first nor, still more gratifying, that a doctoral student produces by his second year already several solid papers and patent applications. The Chinese need no patronizing, least of all by their own newspapermen.

Some Chinese with overseas connection have, by private arrangements, gone to attend foreign universities. Such a self-supporting student is effectively free to stay abroad if residence is not denied by the foreign country. On the other hand, state-sponsored personnel on international exchange 'defect' if for peculiar reasons they refuse to return. Indeed, it is reckoned that since 1979, out of the twenty thousand who have been abroad, only sixteen students and twelve scientists have done so. This number of illegal emigrants is negligible compared to that of overseas Chinese researchers who go to work and often settle in China, motivated by patriotism or higher political ideals.

Apart from the overseas Chinese, foreign scientists are also visiting China in large numbers to exchange expertise or cooperate in research. China led the world in science and technology until the seventeenth century, but was fast outpaced by Europe after the time of Galileo. The lingering economic backwardness coupled with the memory of political humiliation may affect the self-confidence of some Chinese officials in the context of international science. In truth, China has much to contribute in many fields of science and technology. These include acupuncture anaesthaesia and therapeutics, microsurgery, early diagnosis of hepatoma, carcinoma of the oesophagus and cancer of the nose and larynx, genetic breeding of crops and cattles, blast furnace powdered coal injection and dome-combustion hot blast stove, as well as earthquake and flood forecasts. Indeed, since 1974 China has used the funds provided by organizations of the United Nations development system to impart her technical skills to many developing countries, by holding discussion meetings and technical training classes and by dispatching experts. These items include acupuncture, freshwater fish breeding, biogas, silkworm breeding, rural comprehensive development, small hydro-power stations, diesel engine technology and groundwater exploration. Examples of bilateral collaborative research with developed nations are the Sino-French geotectonic survey of the Himalyas region of Xizang and the West German participation in an energy programme for Guangdong.

We now come to examine the development of information management, of which international exchange actually is a part (apart from its bearing on international relationship). The existence of archives and libraries in China can be traced back as early as 3000 years ago or earlier. Most belonged to kings and emperors. The oldest archive still preserved today dates from AD714. In 77BC, at Imperial command, a scholar-minister began to compile, and later his son completed, two pivotal works which laid the seed of information science. One, entitled 'Alternative Register', was a collection of abstracts and the other, 'Summary for Seven Classics', was the first bibliography to be compiled in the world as far as is known. The invention of paper in the first century AD, of printing in the fifth century, and of the moving-block printing technique in the ninth century accelerated the proliferation of books and the rise of private libraries. The publication of the encyclopaedia 'Great Treatise from the Forever Joyous Era', in 22877 volumes plus 60 volumes of index, and the 'Complete Catalogue of Four Treasure Houses' which ran to 36275 volumes, in AD1408 and 1782 respectively, demonstrated the blossoming of information service in feudal

China. Of course, only the rich or the powerful were served. Libraries in the modern sense were products of the reform movement that represented the Chinese reaction to Western and Japanese imperialism. The first provincial public library was founded in 1903 at Wuhan, later the birth place of republican China (where the uprising that overthrew the last royal dynasty was sparked off), and other provinces followed suit.

The first research institute devoted to information science was the Institute of Scientific and Technical Information, set up in Beijing by the Academy of Sciences in 1956 to help its own institutes. (A literal translation of the Chinese title will have 'Intelligence' in place of 'Information'.) Two years later, after the first National Working Conference of Scientific and Technical Information had submitted a consensus proposal to the State Council, the institute was upgraded to be administrated by the State Scientific and Technological Commission directly. It became the national coordinator for information work in all government organs. With its professional advice, soon every ministry, most commissions, all the 21 provinces and the 5 autonomous regions created similar institutes. Furthermore, national and provincial exchange networks began to span the country. There are now about 300 and 3 000 of these networks respectively. Examples are the National Information Exchange Network for Low Voltage Electrical Equipments and Electric Transmission, and the Shipbuilding Information Exchange Network of Guangdong and Guangsi. In addition, centres of scientific and technical information can be found in the seven big cities belonging to the original seven Administrative Regions: Shanghai, Xian, Tianjin, Shengyang, Wuhan, Chengdu and Guangzhou (corresponding to the East, Northwest, North, Northeast, Middle, Southwest and South respectively). At Chongqing in southwestern China the Institute of Scientific and Technical Information has established a branch, best known to the public for its cover-to-cover translation of 'Scientific American'. The number of qualified personnel in all these organizations amounts to over 50 000. Many are members of the China Society of Scientific and Technical Information, the secretariat of which is at the Institute in Beijing.

The common tasks facing the various information services are stipulated in a State Council circular, 'Scheme Concerning the Carrying out of Scientific and Technical Information Work'. They are:
– to provide on request necessary and timely data for the promulgation of policies, objectives, plans and projects of production and research;
– to collect quickly and comprehensively documents concerning important inventions at home and report them as quickly as possible in the form of abstracting journals; to organize the exchange of plans on vital domestic scientific research items and of intermediate scientific research results;
– to compile, translate and report selectively but systematically current scientific and technical literature of all kinds from developed countries, including technical publications, dissertations, conference proceedings, video material, patent specifications, standards, catalogues and samples;
– to effect information analysis; and
– to accumulate step by step and systematically sort out information for the establishment of data bases of different specialities as foundations for

computer retrieval in the future.

Apart from the centres of scientific and technical information, where there are reading rooms open to 'recognized' visitors, the places where primary information in the form of books and periodicals can be accessed are the libraries, which may be divided into five kinds.

1. Public libraries: these include national, provincial or autonomous regional, prefectural and county libraries, down to the district, neighbourhood and street libraries in cities, and in the countryside to commune and brigade libraries. Their governing body is the Administrative Bureau of Library Service under the Ministry of Culture. In practice, however, the libraries are not centrally directed but are largely financed by the cultural bureau of local authorities at the appropriate level. At present, public libraries at county level and above total about 1 800. The largest is the Beijing Library, a national establishment founded in 1910. It is a depository library for all Chinese formal publications with emphasis on social sciences, including 11 000 periodicals, of which four-fifths are from abroad. The second biggest is the one in Shanghai. All these libraries have special significance, as they are within the reach of anyone bent on self-study. A drawback is that they usually operate on a closed shelf basis, with few items besides newspapers on open display.

2. Educational libraries: those in universities and colleges and their constituting departments, and in schools. Their funds derive from general grants made by the Ministry of Education to the institutions they serve. The largest is Beijing University General Library, which has interlibrary loan arrangement with 200 libraries in the country.

3. Research libraries of the Academy of Sciences, the academies under various ministries, and their member institutes. The Academy of Sciences maintains libraries in Beijing, Shanghai, Wuhan and Lanzhou. The biggest is the one in Beijing, with a staff of almost 400 and holdings that include 600 000 volumes of books as well as 750 000 copies of technical reports and material on microfilms. Learned societies also have libraries, but they are usually small.

4. Trade union and factory libraries, which make extensive use of the public, educational and research libraries through group borrowing.

5. Libraries in administrative organizations and army units, functioning in a similar manner to those in factories.

There are 17 universities with departments, specialities or classes of librarianship at present, with 1 500 undergraduates and 1 500 correspondence students. The only course available on information science is in the Department of Librarianship at Wuhan University. The establishment of a college of information science is being considered by the Ministry of Education. As an interim measure, lectures on librarianship and information sciences are given to students in many universities, by professionals from the Institute of Scientific and Technical Information, and the Beijing National, Academy of Sciences and other large libraries. On-the-job training is also provided by these advanced units for the staff from affiliated organizations.

The automation of information processing is under active development. The first success occurred in November 1975 when the Information Institute of the then First Ministry of Machine Building, with help from the

Ministry's Computer Centre, used a China-made DJS-C4 to retrieve 500 foreign papers on metallurgy. In collaboration with the Language Institute of the Academy of Social Sciences and the Statistics Institute of the Academy of Sciences, the Institute of Scientific and Technical Information is experimenting with a software system for English–Chinese titles machine translation. In pilot tests of April 1978, satisfactory results were obtained in translating 20 titles from English into Chinese. This system is still being evaluated on a TK-70 computer for translating papers from English into Chinese both in metallurgy and computer science. In addition to the TK-70, the Institute has a T-4100 machine on which a processing system for information in Chinese characters, imported from Japan, is implemented. Computer-assisted operations in the library are also studied. For the cataloguing and card production of Western bibliographical material, an agreement has been reached among major libraries to adhere to the Anglo-American Cataloguing Rules, Second Edition, with minor variants to suit Chinese practice. Trial runs are being performed on the Felix-3512 installation in the Statistics Institute to create a national union acquisition file.

China now publishes some 7 000 periodicals on science and technology. To list all of them would be out of place in this book. However, the names and addresses of publishers of scientific books and journals are given below.

List of Publishing Houses for Science and Technology

People's Publishing House, 166 Chaoyangmennei Dajie Street, Beijing
Joint Publishing Company, 166 Chaoyangmennei Dajie Street, Beijing
World Knowledge Publishing House, 24 Zhanlanlu Road, Fuchengmennei, Beijing
Workers' Publishing House, Liupukang Andingmenwai, Beijing
China Youth Publishing House, 21, 12th Lane, Dongsi Bei Dajie Street, Beijing
China People's University Publishing House, 39 Haidian Road, Beijing
People's Education Publishing House, 55 Shatan Houjie Street, Beijing
Cartographic Publishing House, 3 Baizhifang Xijie Street, Youanmennei, Beijing
Commercial Press Limited, 36 Wangfujing Street, Beijing
Zhonghua Book Company, 36 Wangfujing Street, Beijing
Youth and Children's Publishing House, 1538 Yanan Xilu Road, Shanghai
Science Press, 137 Chaoyangmennei Dajie Street, Beijing
Machine Building Industry Publishing House, 1 Baiwanzhuang Nanjie Street, Beijing
National Defence Industry Publishing House, PO Box 2819, Beijing
Coal Industry Publishing House, 16 Heping Beilu Road, Beijing
Electric Power Industry Publishing House, Liupukang, Deshengmenwai, Beijing
Petroleum Industry Publishing House, A1 Andingmenwai Guandong Houjie Street, Beijing
Geological Publishing House, 64 Yangshi Dajie Street, Beijing
Surveying and Cartography Publishing House, 50 Sanlihelu Road, Beijing
China Building Industry Publishing House, Baiwanzhuang, Beijing
Textile Industry Publishing House, 3 Fuchenglu Road, Beijing
Light Industry Publishing House, 3 Fuchenglu Road, Beijing
China Railway Publishing House, 14, 3rd Tiao Lane, Dongdan, Beijing
People's Posts and Telecommunications Publishing House, 27 Dong Changanjie Street, Beijing
China Forestry Publishing House, 130 Chaoyangmennei Dajie Street, Beijing
Water Conservancy Printing House, Liupukang, Beijing
People's Medical Publishing House, 10 Tiantan Xili Road, Beijing
Scientific Popularization Publishing House Zizhuyuan Park, Xijiai, Beijing
Metallurgical Industry Publishing House, 74 Dengshikou, Beijing
Chemical Industry Publishing House, No. 16 Building, 7th Qu District, Hepingli, Beijing
Beijing Publishing House, 51 Dong Xinglongjie Street, Beijing
Tianjin People's Publishing House, 124 Chifengdao Road, Tianjin
Shanghai People's Publishing House, 54 Shaoxinglu Road, Shanghai
Hebei People's Publishing House, 19 Beimalu Road, Shijiazhuang
Shanxi People's Publishing House, 7 Bingzhoulu Road, Taiyuan
Inner Mongolia People's Publishing House, 82 Xincheng Xijie Street, Hohhot
Liaoning People's Publishing House, Nanjingjie Street, Shenyang
Jilin People's Publishing House, 102 Stalin Dajie Street, Changchun
Yanbian Education Publishing House, Yanji, Jilin Province
Heilongjiang People's Publishing House, 42 Senlinjie Street, Harbin

Shaanxi People's Publishing House, 131 Bei Dajie, Xian
Gansu People's Publishing House, 230 Qingyanglu Road, Lanzhou
Qinghai People's Publishing House, 76 Xiguan Dajie Street, Xining
Xinjiang People's Publishing House, 306 Jiefanglu Road, Urumqi
Shandong People's Publishing House, Shenli Dajie Street, Jingjiulu Road, Jinan
Jiangsu People's Publishing House, 50 Gaoyunling, Nanjing
Anhui People's Publishing House, 1 Yuejinlu Road, Hefei
Zhejiang People's Publishing House, 196 Changzhenglu Road, Hangzhou
Henan People's Publishing House, 94 Xililu Road, Zhengzhou
Hubei People's Publishing House, Jiefang Dadao Road, Wuhan
Hunan People's Publishing House, 14 Zhanlanguanlu Road, Changsha
Jiangxi People's Publishing House, 3 Baihuazhoulu Road, Nanchang
Guangdong People's Publishing House, Simalu Road, Dashatou, Guangzhou
Guangxi People's Publishing House, 14 Hetilu Road, Nanning
Chongqing Publishing House, 102 Lizibazhengjie, Chongqing
Guizhou People's Publishing House, 5 Yananzhonglu Road, Guiyang
Yunnan People's Publishing House, 100 Shulinjie Street, Kunming
Sichuan People's Publishing House, 3 Yandaojie Street, Chengdu
Shanghai Science and Technology Publishing House, 450 Ruijin Erlu Road, Shanghai
Xinjiang Youth Publishing House, 9 Jianshelu Road, Urumqi
Xinjiang Education Publishing House, 151 Shenglilu Road, Urumqi
East China Normal University Publishing House, 3663 Zhongshan Beilu Road, Shanghai
Yanbian People's Publishing House, Yanji, Jilin Province
Agriculture Publishing House, 130 Chaoyangmennei Dajie Street, Beijing
Shanghai Education Publishing House, 123 Yongfulu Road, Shanghai
Ningxia People's Publishing House, 4 Gongyuanjie Street, Yinchuan
Fujian People's Education Publishing House, 5 Damengshan, Fuzhou
Hongqi Publishing House, 2 Shatan
Inner Mongolia Education Publishing House, 1 Hulunlu Road, Hohhot
Technical Standardisation Publishing House, Sanlihe, Beijing
Xizang People's Publishing House, Jianshelu Road, Lhasa
Fujian People's Publishing House, 27 Deguixiang Lane, Hexilu Road, Fuzhou
Atomic Energy Publishing House, PO Box 2108, Lanzhou
Scientific and Technical Documents Publishing House, Beikou, Hepingli, Beijing
Seismology Publishing House, 63 Fuxinglu Road, Beijing
Guangdong Science and Technology Publishing House, 37 Xinjilu Road, Xiti, Guangzhou
Xinjiang People's Health Publishing House, 76 Longquanjie Street, Urumqi
Yili People's Publishing House, Kuitun, Xinjiang
Shanghai Scientific and Technical Documents Publishing House, 1, 6th Lane, Gaoanlu Road, Shanghai
Oceanographic Publishing House, Fuxingmenwai Dajie, Beijing
Meteorology Publishing House, 46 Baishiqiaolu Road, Beijing
Shandong Science and Technology Publishing House, Shengli Dajie Street, Jingjiulu Road, Jinan
Jiangsu Science and Technology Publishing House, 56 Gaoyunling, Nanjing
China Encylopedia Publishing House, A1 Andingmenwai Guandongjie Street, Beijing
Qinghai Education Publishing House, 106 Xiguan Dajie Street, Xining.
Anhui Science and Technology Publishing House, 1 Yuejinlu Road, Hefei
Bibliographic Publishing House, Beijing Library, 7 Wenjinjie Street, Beijing
Shaanxi Science and Technology Publishing House, 131 Bai Dajie Street, Xian
Hunan Science and Technology Publishing House, 14 Zhanlanguanlu Road, Changsha
Qilu Publishing House, Shengli Dajie Street, Jingjiulu Road, Jinan
Beijing University Publishing House, Beijing University, Haidianqu, Beijing
Metrology Publishing House, 7, 11tBlock, Hepingli, Beijing
Fujian Science and Technology Publishing House, 27 Guixiang Lane, Hexilu Road, Fuzhou
Tianjin Science and Technology Publishing House, 124 Chifengdao Road, Tianjin
Knowledge Publishing House, A1 Andingmenwai Guandongjie Street, Beijing
China Agricultural Machinery Publishing House, 21 Dongdiaoyutai, Haidianqu, Beijing
Heilongjiang Science and Technology Publishing House, 28 Zhenfenbujie Street, Nangang
Zhejiang Science and Technology Publishing House, 196 Wulinlu Road, Hangzhou
Educational Sciences Publishing House, 10 Beihuan Xilu Road, Bei Taipingzhuang, Beijing
Qinghua University Publishing House, Qinghua University, Beijing
China Environmental Sciences Publishing House, Chinese Society of Environmental Protection Xiaozhuang, Haidianqu, Beijing
Patent Literature Publishing House, Balizhuang, Fuchenglu Road, Xichengqu, Beijing
Beijing Normal University Publishing House, Beijing Normal University, Xiaoxitian, Beijing
Astronavigation Publishing House, PO Box 849, 24, Beijing
Henan Science and Technology Publishing House, 94 Xililu Road, Zhengzhou
Industry and Commerce Publishing House, 45 Fuxingmennei Dajie, Beijing
Sichuan Youth and Children's Publishing House, 3 Yandaojie Street, Chengdu
Ancient Books of Traditional Chinese Medicine Publishing House, 11 Beihuan Donglu Road, Beijing
Guangdong Provincial Cartographic Publishing House, 468 Huanshilu Road, Guangzhou
Fujian Cartographic Publishing House, 1 Wenquanlu Road, Fuzhou
China Science and Technology Publishing House, Zizhuyuan Park, Xijiao, Beijing
Fudan University Publishing House, 220 Handanlu Road, Shanghai
Central China Polytechnical University Publishing House, Central China Polytechnical University, Yujiashan, Wuchang
West Lake Calligraphy Publishing House, 196 Changzhenglu Road, Hangzhou
China Academic Publishing House, 137 Chaoyangmennei Dajie Street, Beijing
Xinjiang Kashi Uygur Publishing House, Kashi, Xinjiang
Printing Industry Publishing House, 2 Cuiweilu Road, Fuxingmenwai, Beijing
Rural Publications Publishing House, 61 Fuxinglu Road, Haidianqu, Beijing

2 Science and Social Change in China*

2.1 Nature and Dynamics of Chinese Society

Chinese society may be thought of as having a triple nature, or to put it in a different way, to evince three tendencies. To facilitate explanation, they will be referred to as the three social formations, although they are meant, each, to include the whole population.

The first of these formations, which comes nearest to traditional Chinese society, is referred to as the bureaucratic formation. (The word bureaucratic is not used pejoratively.) The second formation, which comes nearest to the nature of the advanced Western countries, is referred to as the legalist formation. The term is appropriate because of the importance of the concept of law, both human and natural, and also because it can be closely related to the ancient Chinese school of thought known as the Legalists. The third is the revolutionary formation, because it is particularly involved with rebellion and social change.

The aims of this chapter are to outline the three formations, and to suggest how they relate together and interrelate with the development of science.

Firstly, then, a consideration of Chinese society as a bureaucracy administering an atomized people. The bureaucratic formation is divided into three sections, the executive, the officials and the people. It is administered partly by a code and partly by the arbitrary decisions of the top executive, and at each level below, the arbitrary decisions of the officials at that level and those immediately above them. It is characteristic, that so long as the people abide by the instructions and decrees of the officials, they are left to carry out their work and live their lives unmolested.

Formally in China, the executive can be thought of as the leadership of the Communist Party (Party); the bureaucracy as composed of those who are referred to as cadres and paid on a cadre scale; and the people as all those who are paid a wage, those who receive pocket money (students, and soldiers) and all the peasants, except those who have been appointed from above to administrative positions.

The bureaucratic formation includes the following features: the life of the whole people follows a harmonious and cooperative routine, which is felt to be right and proper and in accordance with the common good. All those in superior positions are as such respected and feared; their favour is courted,

*By John Collier (Scottish-Chinese Friendship Association, Edinburgh, UK)

their power resented, and overall, contact with them is avoided by all beneath them, except where specific support is required. Because power is wielded arbitrarily, every group or grouping tends to hold together, while at the same time keeping aloof from other groups. This tendency runs right through society from the individual and the family, right up to whole provinces and regions of the country, and the top executive. Every bureaucratic centre, whether it is an individual official, or a committee or organization, endeavours to extend its power, but this endeavour is primarily defensive, and is aimed at quantitative growth not qualitative change. Thus its all pervasive characteristic is a conservative spirit of live and let live.

Next we come to consider society as a syndrome of citizen, property, law and constitutional government. The legalist formation divides the Chinese people into four parts: the Party executive, the government officials, those with professional status, and finally all those without professional status or office.

Whereas in its role as executive to the bureaucracy, the Party leadership is legitimized as architect of the plan, protector of the people and provider of social initiative, its role in relation to the legalist formation is akin to the apex of constitutional government, formulating the principles which will govern the plan, protecting property and law and order, promoting democratic procedures, and guaranteeing professional authority and status.

The legalist ethos emphasizes the following factors: people as competing individuals, who have the right to be paid according to their work, and to have authority and status in accordance with their responsibility, their social contributions and their past educational and professional achievements. It encourages people to pursue their own and their family interests, and to be ambitious for more pay and promotion, and to register individual and group achievements socially. In short this ethos may be likened to the prevailing middle-class ethos in Britain.

Thirdly and lastly we consider society in its involvement in mass campaigns and political struggle resulting in social change. This, the revolutionary formation, precisely because it is society in a dynamic condition of struggle and change, is less easy to define further. Here, it is less easy to divide the population into definably different sections. It certainly arises from the division of society between those with authority and those without it. However, the immediate dynamic division is between social activists and those who are drawn into mass-activity, on the one hand, and those who cling to and defend the old authority, whether official or professional, on the other. There is also the complication, that when there is political polarization and conflict, there will be activists on both sides, as was the case in China in 1966 to 1969. That is both rebel and conservative activists.

The ethos of the revolutionary formation is that of equality and leadership, friend and foe, and the validity of social change. It constitutes a threat to all authority and status, and an opportunity for all those who previously felt oppressed or exploited.

Below, it will be shown that each social formation has roots in Chinese history, and is also integral to her era of turmoil, which commenced early in the nineteenth century and may be considered to be still in progress.

It is unnecessary to emphasize the Chinese experience of bureaucracy, as to many people China is the archetype of such a society. What may need emphasizing is the positive aspects of this society. During the two millenia of the Empire, China enjoyed a preponderance of years of peace, prosperity and cultural development, which have been compared by many, very favourably, with the same period of European history. Furthermore, during the first three-quarters of this epoch, China led the world in technology, and crucial proto-scientific discoveries. Thus aspects of the bureaucratic tradition are peacefulness, tolerance and a minimum of interference in community life by government, either through law or directive.

The legalist tradition goes back to Confucian times. The Legalist School was peaceable, preaching a defensive stance, if not outright pacifism. It was concerned with technology, with individual merit as against inherited or official status, and with rational administration. In fact it has been likened in its prescriptions to the Napoleonic Code. Although it was favoured by the first emperor, it subsequently lost ground to both Confucianism – which favoured conservative bureaucracy – and Taoism, which favoured quietism and anarchy. However, Legalist ideas continued in academic studies into language and history, and continued to be fuelled by commerce, industry and urban administration.

With China's defeats in the Opium Wars, many Chinese adopted the view that however superior China was ethically, Europe was superior in power; and this power was associated with European education and culture. Thus together with the physical and cultural penetration of China by foreign powers, there grew up a strong Chinese urge to learn from them. Representative and symbolic of this, the founder of modern China – revered both in Taiwan and in the People's Republic – Dr Sun Yatsen, had a Western schooling and medical education, and lived a great part of his life abroad, being strongly influenced by the social and political thought of the West. Since the May Fourth Movement after the First World War, China's Renaissance, scientific education has been strongly fostered and extolled, not only as the basis for material progress, but also as the antidote to debilitating superstition. This can only enhance the legalist ethos.

China maintained throughout the imperial era the doctrine of the Heavenly mandate. Part of this doctrine holds that whatever the cause – natural disaster, inefficiency, corruption or foreign invasion – if the Emperor fails to maintain stability and the prosperity of the people, then he has erred and Heaven is angered. Should he then be overthrown, this is proof that he no longer has the mandate of Heaven, which has been transferred to his vanquisher. Thus rebellion is justified, if it succeeds. Together with this doctrine, the authoritarianism of Confucianism was balanced by Taoism, which, if it did not necessarily support rebellion, mocked at power and was subversive of authority. It may have encouraged a passive quietist response to natural disaster or human harshness, as did Buddhism, up to a point; beyond this point it accepted chaos, anarchy and rebellion.

The last great peasant rebellion, which engulfed a large part of China for fifteen years from 1850, was partly inspired by Christian doctrine, and embraced the concept of social equality, including the emancipation of women, and common property. Again, furthering the ethos of struggle and social change, the October Revolution in Russia, not only led to the creation of the Chinese Communist Party, but deeply affected the whole Nationalist movement. The revolution was seen as the first great break with European imperialism, which was paralysing China. From its foundation to its coming to power, the Chinese Party was inspired by the Russian Revolution, and learnt from it. But not primarily from the bureaucracy and authoritarianism of Russian life in the Thirties, but rather the revolutionary seizure of power of 1917, and the social transformations that followed it. This became reinforced, because from 1925 to 1949 the communist-led areas of China relied for their survival on the positive support of the peasants, and this support was forthcoming, not so much in the spirit of Confucian submission to authority, as in the spirit of rebellion against a failing emperor, and of preparedness to fight a foreign invader.

2.2   The First Seventeen Years of the People's Republic

So it was then, that the People's Republic came into existence with all three of the social formations well entrenched. Both in the establishment of the People's Republic and its evolution since 1949, the three formations have all been strongly represented, and have interacted together. Each formation has had periods of growing strength and influence, and periods of partial decline and eclipse.

The last years of the Nationalist Government saw the democratic parties and their leaders attacked and persecuted by the government, widespread corruption and a galloping inflation, which brought ruin to wide sections of the middle classes. Furthermore the communist leadership was widely respected for having honourably fought against the Japanese invaders, and for honestly representing the interests of the poor people of China, both workers and peasants. The country was in ruins, and the Communist Party was relatively small and inexperienced in the face of the enormous tasks of stabilization and reconstruction. In these conditions there was a basis and a dire need for cooperation on the part of all sections of the people – not least the army of petty officials who had had posts in the organs of the Nationalist bureaucracy.

The constitution of the new republic was created with the positive involvement of the democratic parties, and non-party middle class, through the People's Consultative Congress. The resulting constitution, as well as recognizing the leading role of the Communist Party, brought into being a new constitutional structure of the European democratic type, however limited its powers. The needs of reconstruction, coupled with the extreme shortage of trained personnel, meant a renewed status for those with education and advanced training, and all aspects of the development of science and technology became high priorities. Thus both the tradition of bureaucracy and the new legalism got off to a good start after 1949. In the early years, four mass-movements assured the continuation of the revolutionary momentum. These were: an anti-rightist movement, which was intensified by the outbreak of the Korean War; two campaigns against corruption in industry and the civil service; and the completion of land reform, which had been commenced in the north in 1946.

From 1952 to 1957 two main developments took place. Firstly, the countryside experienced a number of mass-movements, which culminated in the great majority of the peasantry and the ex-landlords ending up in large cooperatives, in which people were paid more or less in accordance with their work, and nearly all farm capital was owned collectively by the cooperative members. Here then the revolutionary formation was dominant, both as a mass movement and as social change. Secondly, the rest of society was taken up with the First Five Year Plan, in which the Soviet Union was taken as a model, and Soviet advice, guidance and material assistance was deeply involved. Russian became the first foreign language, and tens of thousands of Chinese were educated and trained by Russians. Because of the authoritarian bureaucratic nature of the Soviet Union in the Fifties, together with the professionalism of the many Soviet advisers and educators, Soviet influence must have both strengthened the authoritarian and status-conscious aspect of Chinese bureaucracy, while at the same time strengthening professionalism, and inculcating it into the new generation of Chinese graduates.

In 1956, a mass-campaign primarily concerning private business interests virtually brought the private employment of labour to an end in the cities. This transformed the private employer into a salaried manager, under a joint state-private board, albeit still receiving a fixed return on his capital.

The period may be summarized as follows: in the countryside, the momentum of revolution was maintained, but at the same time through the new constitution, the spread of education and the planned procurement of food and raw material for the cities, together with the extention of water control, bureaucracy and legalism were also developed. The city ethos of the period favoured bureaucracy and legalism, but there were also mass-campaigns, particularly concerning public health, as well as the campaign to transform private business already mentioned.

The following period, which has come to be referred to as that of the Great Leap Forward was very complicated, and fraught with twists and turns, and conflicting assessments; but overall can be categorized as one in which both social change, and the ethos of social change, were dominant. However, the failures and material losses led to the strengthening of the other two formations. In particular, they brought home the need for careful planning, the importance of scientifically based technology, and the value of balanced economic development.

The period 1961–66 was one of emphasis on planned production and professionalism; raised output, but also raised standards. However, it was also a period of social experiment, and political and ideological struggle within the Party. Both aspects can be seen in the opening of the Daqing Oilfield (Chapter 7.2). This development on virgin land in the northeast, involved mass-campaigns of voluntary work, and new institutions. Instead of building a large urban technical and administrative centre, emphasis was put on carrying out such work on the work site, and a number of residential villages were constructed, combining industrial and agricultural workers. Instead of importing food for the oil workers, or having a parallel development of agriculture in the area, food production was developed – mainly by women – by the oil workers' family members, with the conception of uniting industry and agriculture in one community. The village concept, together with the spartan living and working conditions of all personnel, created a strong egalitarian tendency, in advance of what was happening in the country generally.

In 1962–63 a mass-campaign was launched, called the Socialist Education Movement (SEM). This was aimed primarily at various forms of corruption, and the pursuit of narrow individual interests. The ideological dispute had as a major focus the diametrically opposed views, that on the one hand class struggle was alien to socialism, and in fact dying out in China; and on the other, that it was integral with socialism, and would wax and wane but not die out for a protracted period. The public argument was carried out in obstruse philosophical terms but its importance can be judged by the fact that it was a central controversy underlying the Cultural Revolution, and recently the denigration of that movement.

The SEM was of particular interest in relation to the three formations. Two factions arose in the Communist Party, one of which was in a sense legalist, the other revolutionary, while both criticized the other for being bureaucratic. The faction led by Mao Zedong considered the SEM to be a political movement to advance socialism by arousing the peasants and workers to criticize those in authority who were corrupt or self-seeking, and through this criticism to come to an awareness of their own shortcomings. The faction led by Liu Shaoqi, and the present leader, Deng Xiaoping, saw the movement more in rational, legalistic terms, requiring that all corrupt persons be criticized and penalized, in the form of correction from above.

2.3   The Cultural Revolution and its Aftermath

Mao Zedong forecast that the Cultural Revolution would go through three stages, and this was encapsulated into three Chinese characters, which mean roughly, 'struggle, criticism and transformation'. This is a reasonable way to characterize the three years commencing in the summer of 1966. The first saw the power structure and the organization of the Communist Party as it existed in 1965, and much of the administration of the country, destroyed or disrupted. The second year saw the country divided up into innumerable separate groups and groupings, and overall into two hostile camps. These engaged in an outpouring of criticism, which ranged from self-criticism at one pole to hostile condemnation leading to violent struggle at the other. The third year saw the re-emergence of the structured power and organization of the Party, the government and the armed forces, and the implementation of a number of major social reforms. The period from 1969 to 1978 was one of emphasis on construction, but this was inhibited by a condition of intense factional struggle, which went through a series of changed alignments.

The intellectuals were comparatively favoured from 1960 to 1965, and have been again since the renewed legalistic leadership gained predominance in 1977–78. On this basis the picture now presented by the leadership in China seems very simple. It asserts that the legalistic criticisms of Mao and the revolutionary policies of the Great Leap, which were made by many intellectuals in 1960–61, were justified. The support given to the intellectuals in the period up to the Cultural Revolution was correct, but marred by a continued leftist line pursued by Mao. The intellectuals were hounded during the 'ten bad years' by the Left with the support of Mao and the ameliorating influence of Premier Zhou Enlai, and since 1977, when Deng Xiaoping gained ascendancy, their position has steadily improved. This picture is in some respects misleading, and the following account is more in accordance with the facts, and more relevant to an understanding of the relationship between science and social change.

The Cultural Revolution, which had been initiated in 1963 as a parallel movement to the SEM in the sphere of culture, was originally led by Peng Chen, then Mayor of Beijing. Mao saw it as a movement of criticism of the old academic authorities from below, aimed at bureaucracy and the failure to stimulate a new socialist culture. Peng and the organizational leadership saw it as a more formal self-criticism movement, led by the Party. In the spring of 1966 Peng was overthrown, and a new Cultural Revolution Committee was formed. This committee eventually evolved into the Leading Group, which was finally itself overthrown in 1976 and imprisoned as 'The Gang of Four'.

The Cultural Revolution Committee held the initiative in the media, and encouraged criticism from below. When this developed into criticism of Party committees at various levels, the organizational leadership appointed working parties which were sent into urban institutions, particularly important schools and colleges; and the initial criticism was suppressed. At the same time the discontent of the students was turned against low-level officials, lecturers and teachers. To counter this, Mao initiated the Red Guard movement, which by August spread throughout the country. In the name of Mao and socialist progress, the Red Guards attacked on a broad front. They searched the houses of pre-liberation property-owning and official families, and destroyed or confiscated what they considered bad or illegally held property. They held meetings of denunciation against those they labelled reactionary scholars and teachers, and they defaced or destroyed what they termed feudal relics in public places.

However, the Red Guards were limited to the so-called 'Five Red Categories' – children of workers, peasants, armymen, martyrs and communist cadres – and they tended to come under the leadership of the sons and daughters of leading cadres. They also split between those who supported the first rebels, and those who supported the working parties appointed by the Party. The groups that gained an ascendancy in the autumn of 1966 were mainly led by children of the top leaders at each level, most of whom had been criticized. So they were strongly influenced by the Party organization, and focused their attacks on the old intellectuals, the Cultural Revolution Committee and the initial rebel students, while defending the old organizational leadership.

In the winter of 1966, in order to counter the influence of the Party organization, the Red Guards were thrown open to all, and the top organizational leadership was overthrown. The new entrants to the Red Guards were largely from professional families, and had the support of their parents to join the rebel side. This was particularly the case because the Cultural Revolution Committee, wishing to direct the attack against the Party leaders, attempted to direct it away from the intellectuals. From this time (1967), because of these developments, the majority of professional families, both students and their parents, adhered to the rebel side led by the Cultural Revolution Committee, until many became disillusioned in the 1970s. This tendency was partly sustained due to political commitment and respect for Mao Zedong, and partly because the rebel students had reason to believe that if the old leaders were cleared out they would qualify for professional advancement, and be the natural successors to political leadership and office.

In 1968 the policy was adopted of encouraging all the Red Guards and other school and college graduates, to go to live and work in the countryside: some for a few years, many of them to live permanently. At the same time, as well as sending those cadres who were considered to be rightists to the countryside for prolonged periods, the policy was adopted that all cadres (all those in positions of authority) and clerical workers should go to the countryside for periods ranging from three months to one year, and this should be continued in the future. Most medical graduates were assigned to practise in the countryside, and many colleges were dispersed into the interior, away from the eastern cities; while college work was integrated with industrial and other work. Groups of workers were sent into the schools and colleges to supervise and assist with administration. This was intended to break down academic attitudes and the ethos of exclusiveness. It was decided that from this time on students for higher education should only be recruited from among people who had already worked for some years in industry or agriculture, and that they should first be recommended by their workmates. Many scientific research institutes remained closed or were starved of funds, and many were dispersed into the interior. The physical sciences faired better than most, but were closely tied to industry, and the needs of particular enterprises.

All these measures were damaging to professionalism, and reduced the status of the intellectuals, including the scientists. Furthermore in the years of intense factional struggle that followed, the leftist restrictions on culture in general were only gradually relaxed, and great emphasis remained on schools and colleges producing their own curricula suited to local conditions, and on art and literature of workers and peasants, as opposed to professional writers and artists.

Before summing up the position in 1978, a further important factor should be mentioned. From 1956, when the Soviet Union refused to support

China over the Taiwan Strait issue, relations between the two countries had deteriorated, both politically and ideologically, until they reached outright hostility with the military incidents on their borders in 1969. With China wishing to concentrate on a modernization programme, it was clearly unsatisfactory to have bad relations with the Soviet Union and the West and Japan. Furthermore, whereas in 1956 with the launching of the Sputnik it seemed that Soviet technology was overtaking the West, in 1969 the situation looked very different, with America and Japan appearing to be in the lead. With the end of the Vietnam War, Mao took the initiative to heal the breach with the United States, and large-scale trade and commercial relations were initiated.

Just as the political and ideological alignment with the Soviet Union had facilitated commercial and cultural cooperation, so it was clear that if China was to benefit to the maximum extent possible from cooperation with the advanced capitalist countries, professional liberalization would greatly facilitate the process.

In 1976 the factional struggle came to a head, probably precipitated by the death of Premier Zhou Enlai. In his last year, Zhou probably favoured Deng Xiaoping, and was coming under the fire of the Left. However, his commanding position in the government meant he both bridged the gap between Deng and Mao, and was indispensible to stable progress. In April 1966 Deng was again banished from office, and the leftist leaders held the initiative with the support of Mao. However, by 1976 large numbers of old cadres had been brought back into power, especially in 1975, so that the Party balance had tipped back to the Right. The *coup de grâce* to the Left came, when after Mao's death, the new leadership split, and the so-called Gang of Four were arrested on the initiative of Premier Hua Guofeng. The rump of the leftist leadership steadily lost influence until its virtual eclipse in 1978.

The position in 1978 can be summed up as follows: the governing leadership was predominantly the organizational leadership of the Party of 1965, but with its policies modified by the exposure – brought about by the Cultural Revolution – of the stultifying affect of bureaucracy, the strong ethos of liberation among the young, and the liberalizing effect of much greater contact with the rest of the world, dating from 1971 (Deng, for example had represented China at the UN in New York in 1973). The people in general were not only deeply affected by the Cultural Revolution, but had also become politically apathetic and alienated from the Party, due to the subsequent factional struggles at the top. It was thus prone to revert to traditional attitudes, and to put individual and family interests above the collective interest. The intellectuals in general, although the scientists to a lesser extent than others, had come through a tough time. Thus they were suspicious, withdrawn and very much on their guard. Overall, nearly everyone had been associated with one side or the other, and one or more factions. Thus the population was deeply divided. The old leadership was in the majority, but there was tens of thousands of new cadres, who had benefited from the Left's ascendancy in the previous ten years; although key figures might be purged or demoted, the bulk of them would remain.

The leadership of 1978 may be characterized as legalist, but to consolidate its power it had to come to terms with the bureaucratically inclined administration. To create an ethos of continuity, which would legitimize its own position, it had to come to terms with Mao Zedong's thought, as expressed in the five volumes of his collected works.

The expressed aims of the new leadership are the 'Four Modernizations'. To achieve these, there are four basic requirements: the hard work of the whole population, the enthusiastic involvement of the professional experts, the unity of the whole people, and the commercial and scientific cooperation of the most advanced countries. All these things can best be achieved by adopting a strongly legalist policy. The population, which was alienated from the political leadership, will only respond to the enticements of material and cultural benefits to the individual, the family and the local community. The intellectuals will only respond to guarantees of future safety and status, and the offer of what they consider to be appropriate employment here and now. The advanced capitalist countries are encouraged by the prospect of a non-revolutionary stability, and are more easily dealt with in an atmosphere of liberal commercialism. Thus, the leadership, the needs of the country, and the inclinations of the people in general, all give support to a legalist ascendancy at the present time.

This conclusion might have been drawn in one short paragraph. The reason for the outline of past Chinese development, in this and previous sections, is to suggest that this ascendancy, however strong it may be at present, should not be thought of as necessarily stable or permanent.

There is one other major factor which favours continued political struggle and social change. China, for all its unifying cultural tradition and its powerful centralized state and party, is basically a plurality of regions and centres of power. In this respect it is more akin to Western Europe, or perhaps the European Economic Community, than it is to a centralized power such as the Soviet Union. The major regions are the following: the river basin region of the Northern Plain, converging on Tianjin and the Beijing-Tianjin axis; the lower Yangtse basin, centring on the Shanghai-Nanjing axis; the North East Plain with the Dalian-Shenyang-Changsha-Harbin backbone of cities running up the middle; the Northern Plain including the Yellow River basin, with the Taiyuan-Xian-Jinan triangle encompassing it; Southern China, including the Pearl River delta, and the Wuhan-Guangzhou axis; and finally, Sichuan Province including the upper Yangtse basin, and the Chongjing-Chengdu axis. These regions are not definitive, as they may overlap, subdivide or form unions, but they are all self-sufficient, have large urban centres, have exerted independent initiative in the last hundred years, and, perhaps most important, have experienced a different history, both during imperial times and in the twentieth century.

2.4 Science and its Expression in China

Since the Renaissance, the rise of science in Europe has been associated with the development of individualism, rationalism and the quest for objectivity.

The modern Western scientist tends, therefore, to be an observer-manipulator. In his or her view the outside world is comprehended by the mind, which through brain and hand manipulates, in turn, the outside world. Through experiment and theory, these two activities are brought more and more into conformity, and become more and more extended. The perspective is of mankind subduing and harnessing nature to its needs, while achieving a definitive understanding. The focus is on the individual scientist and his or her work – a Newton, Darwin, Mme Curie or Einstein, it is seldom upon the whole structure of scientific theory and endeavour, and less still upon the whole experience of mankind out of which science has arisen. The typical scientist of today is not Renaissance man – with many-sided interests in all forms of knowledge – but a specialist.

The Chinese tradition is very different. It does not strive to subdue nature, nor to reach understanding through analysis. The aim is to come to terms with nature, to achieve and sustain harmony. The men of learning aimed at a wisdom, combining the subjective and the objective, that would allow them to guide the people in harmony with nature. The dictum, 'Dig the channels deep; keep the dykes low' was not just good hydraulic practice, it was also a philosophic plea for harmony and minimum interference. The Chinese tradition, being both wholistic and humanistic, could not separate science from ethics or aesthetics, the reasonable from the good or the beautiful.

The ethos of harmony with nature also involved the cooperation of people. An ambition of the Chinese peasant may have been to work alone with his family on his plot of land; that of the scholar to sit alone in his study writing poetry and painting; but Chinese geography and climate dictated that again and again in life people were induced to cooperate with large numbers of other people, to build and repair dykes, to fight flood, drought, earthquake and insect plague, and when their villages were washed away, to return and rebuild them. And even in peaceful and prosperous times, the irrigation water had to be controlled and shared out, and the tax grain shipped to the great granaries. The Chinese achievement is not based so much on a puritanical work ethic, as on a genius for administration and cooperative endeavour, which are not just two separate abilities, but two aspects of a unified culture.

There is a tendency in China, when a technological problem crops up, to attempt to solve it by a collective and empirical approach. A group of people will patiently carry out an unlimited number of experiments and trials until the problem is solved. This is often the case, when one person with a sufficient theoretical understanding, could have solved the problem more quickly and economically.

It is interesting to note, that both during the Cultural Revolution and in the work responsibility policies of recent years, there has been an emphasis on the individual. In the Cultural Revolution people were encouraged to stand on political principal, so that even if it meant one person standing alone against a large group, this was the order of the day. Under the new policies, both work and payment are arranged so that each person should be encouraged to take personal responsibility.

It is probable that an increased readiness to apply theory and to cultivate individual innovative initiative will benefit Chinese science and technology, but it would be a mistake to undervalue the wholistic and collective tradition. Today much advanced research requires the assembly of a vast and complex array of personnel and equipment, and thereafter its harmonious deployment. The Chinese are still demonstrating that they are particularly good at such undertakings, and they would surely suffer a loss if they became less able in this direction, due to excessive individualization. Again, in the West, the concept of the scientist as a passive observer of nature, merely reflecting what happens 'out there', inhibited the study of mind and brain. Once the idea that consciousness and concept were aspects of a single process, involving the brain and the source of stimulus these areas of study were opened up. As computers develop from being machines that are turned on and off and are restricted to defined processes, to becoming continuous processes in themselves, which develop in interaction with other computers and human beings, the wholistic turn of mind may be expected to come more into its own.

Further, in relation to the form of scientific endeavour, there are indications that the age of extreme specialization may be coming to its end. This, if it is the case, may have a number of implications. Firstly, education may become less specialized, with greater emphasis on acquiring skills and attitudes, which have a wide or even universal application; rather than specialist skills and specific knowledge. Secondly, with education orientated to data retrieval, problem solving, creativity and cooperative work, there is likely to be an increased mobility in work, coupled with a greater degree of interdisciplinary and multidisciplinary research and development. Thirdly, there may be a breakdown of the hierarchic structure of the division of labour, where there is unskilled repetitive work at one end and large-scale centralized decision making and problem solving at the other. Today's science and technology has created the basis for centralized large-scale mass-production and professional exclusiveness; but it has also equally created the possibility of viable small communities, creating and satisfying their own material and spiritual needs in more flexible and mutually participating ways (cf p193).

Finally, in relation to the Chinese wholistic and collectivist tradition and practice, it is relevant to consider the question of the growing menace of pollution and the importance of ecology. It is probable that in the advanced Western countries, the shortcomings in relation to pollution, the use of limited natural resources and the destruction of natural amenities, have not been solely due to the pursuit of private profit and the lack until recently of ecology as a scientific discipline, but has also been due to the dominance of a legalistic rationalism − to an analytical as opposed to a wholistic turn of mind. China has erred in respect of pollution, at least until recently, as badly as most other countries. However, with such pressing problems of poverty and backwardness as existed in 1949, it is not surprising that ecology was neglected. Now that the subject has been brought to the fore there are many indications of rapid progress (Chapter 4.5), and in spite of some signs that the present legalistic ascendancy, with its emphasis on individual competition and self-betterment, is to some extent inhibiting the process, there is reason to be optimistic for the future.

2.5   Immediate and Longer-Term Perspectives

China is too diverse to describe fully its social condition in a few pages. All that can be done here in order to create a basis for considering possible directions of change, is to identify certain important phenomena which concern majorities of the different sections of the people: the peasants, the workers, the soldiers and the students and professional people, which include the scientific workers. These sections of the people and their areas of work will be discussed below in turn.

Until recently most peasants worked in groups. The production team tended to assemble together in the morning, and again after the midday meal before setting off to work in the fields. Work was carried out under the direction of the elected team leader. With group productive work, many meetings and many assemblies connected to particular campaigns, the collective ethos in daily life was very marked.

Today although the team still supervises production plans, and much work is still collective, such as irrigation work and some harvesting, most land is allocated to individual families for continuous cultivation, and their work is paid for according to output. Public affairs are now dealt with by the township administration, which has been separated from the commune – now essentially an economic unit only. The peasants now tend to work alone or in family groups. The egalitarian ethos is discouraged, and people are positively encouraged to get rich through hard work, innovation and enterprise. Television, as well as bringing the village closer to outside affairs, is bringing popular light entertainment to the peasants. Overall, therefore, there is a trend towards a more family and home-centred existence, and away from an involvement in collective activity and public affairs.

Traditionally a major focus of peasant life was the rearing of a large family of sons. The national policy today is for one-child families, and there is considerable material and social incentive to 'encourage' compliance. The political adherence to this policy, let alone its fulfilment, must inevitably involve a marked change in the social conditions of the village. With around a quarter of families with only girl children, the economic and cultural condition of women must also undergo marked change.

With increasing mechanization of agriculture, and diversification of the village and small-town economy, scientific education will increase in the countryside. The reduced family size will in the long run be conducive to parents striving even harder to get their children well educated, and qualified in technical work.

Most of what has just been said about the peasants applies in a modified form to urban wage earners. However, there are some marked differences. In the past most workers were paid on an eight-grade flat time-rate scale, and most promotion to a higher grade went with seniority. Also the focus of the production effort was the fulfilment of shop and factory quotas. Thus there was an emphasis on the collective, and this was reinforced by periodic mass-campaigns. Today work is organized where possible so as to put responsibility for economy and safety, as well as quantity and quality, on the individual worker. At the same time piece-rates and individual bonuses are strongly encouraged, and an increasing premium is put on skilled work.

The accelerated development of science and scientific education in the coming years seems assured. However, their lateral spread seems more uncertain. There are several reasons for this. In the past there was a considerable degree of diversity within a particular productive enterprise. This took several forms. A plant might make some of its own machine tools, as well as a wide variety of products. It would normally also engage in some research and development. There was usually a high degree of vertical and horizontal integration, and an expanding enterprise might well design, build and equip its own extensions. Furthermore, every region, city and even commune, aimed at self-sufficiency by producing a wide range of industrial products, as well as food, raw material and energy. All these practices encouraged a broad spread of local initiative and innovation, which in turn encouraged a wide spread of scientific education and training.

These practices have not been eliminated, but they are being modified with the present emphasis on the economies of specialization. This includes professional specialization, and also plant, enterprise and regional specialization.

Throughout the world the view has been expressed, that the present enormous expenditure on military research and development has restricted civil research and held back theoretical advance. This view may hold true for China to some degree, but in her case there are some important special features. Firstly, the Chinese army is involved a great deal in civil construction (Chapter 8.1). Secondly, to the extent that military R&D are integrated with civil programmes, the security factor is less inhibiting. Thirdly, the proportional level of the military budget is lower than in most countries. Her defence policy, which is partly based upon air-raid shelters and a civilian militia, and partly on a cellular system of self-sufficient communes, allows this.

Since 1949 the Chinese army has been developed partly as a school of education and training for civilian life. After three to five years of service, the mainly peasant recruits have gone back to their villages and towns to take up skilled jobs and positions of leadership. The raising of the technical level of the armed forces therefore raises the level in the countryside, and supports diversification.

Turning finally to professional workers and students. From the setting up of the People's Republic until today, certain conditions in which they have lived and worked are in marked contrast to those of their counterparts in most other countries. Living and working conditions have been exceedingly simple and restricted: the scientists' basic and much-shared laboratory equipment; students in all colleges living eight to a dormitory, with barely enough room for four bunk beds, one long table and benches to sit on; most families with three generations in two or three rooms, with one water tap and a simple cooking stove; libraries packed with students shoulder to shoulder, and usually an inadequate selection of books; nearly all, young and old, with just the bare minimum of food, clothing and shelter. Thus in general professional people including the academics shared the spartan life of the whole population.

The emphasis on science and technology has not just been a government policy, but a widely felt need for improving the life of the people. Partly due to this acute need for progress, and partly due to the condition of bureaucracy, people have tended to concentrate in small groups on their own tasks. However, the intellectuals have also been involved in all the mass-campaigns, from the extermination of flies and backyard steel making, to land reform and the political radicalism of the Cultural Revolution. Furthermore, academics together with other professional people have been allotted a key role as expositors and leaders in the campaigns.

The intellectuals and scientists have been treated as a separate group, separate from the workers and peasants, and separate from the bureaucracy, although quite a high proportion of professional people are Party members, and all have some degree of bureaucratic authority. As a separate group they have been revered and encouraged as the carriers of China's cultural heritage, creators of the new socialist culture, and as key workers in raising production and living standards. As against this, they, and particularly those educated in the old society, have been criticized and repressed as per-petuators of the old elitism and oppressive culture. The scientists have in the main experienced support, but they also have experienced hostile treatment. Because of these particular conditions and circumstances, professional people in China evince a curious contradiction. On the one hand they have been closely integrated with the whole people, and dedicated in their work for material and cultural progress; on the other hand they are now quite isolated, inward looking and taken up with their own narrow self and family interests.

At the present time, the conditions for China's research work and research personnel are improving in nearly all respects. Relatively and absolutely, a growing quantity of resources are being devoted to research. Increased emphasis is being put on scientific work and scientific theory. The status of professional people in general is being raised. Chinese postgraduate studies abroad are being sustained. The supply of research equipment is being upgraded and increased, including the import of advanced equipment. Joint projects with institutions in other countries are growing in number, and routine communication is improving. Scientific work with UNESCO and other UN agencies is being extended rapidly. The number of scientific institutions and journals is growing in response to the ever growing need. In 1982 a move was started to broaden the curriculum of science students, and to include optional subjects including those in the arts. Finally, the working and living conditions of professional people are being improved, and in some cases the burden of teaching is being reduced in favour of extra time for research.

The whole movement of science and technology in China at present would seem to favour a rapid and sustained strengthening of the legalist formation, which up to now, in a sense, has always been manipulated by the other two formations. This manipulation has been exemplified by arbitrary changes of policy towards constitutional and professional institutions and their personnel, on the part of the Communist Party and the government. In the future, this will be less easy, as the legalist formation becomes stronger, more autonomous and integrated into the lives and culture of the people.

A crucial question for the future is whether the revolutionary formation will survive and re-emerge, or whether it will be suppressed for a prolonged period of time. If it is sustained, it is probable that the organization and balance of scientific endeavour in China will differ considerably, both from that in the West and that in the Soviet Union. It is likely that there will be more emphasis on all sections of the people having being involved in science and innovation generally, and it is also likely that scientific institutions will be caught up again in political struggles and movements for social change, as they were in 1950, in 1958 and more radically and disruptively in 1966.

If China experiences a long period of political stability, there is a possibility that it will become more bureaucratic, particularly in the form of institutions becoming dominated by their administrators. This may have an adverse effect on all forms of innovation including scientific research, as seems to have been the case, and still may be the case, in the Soviet Union.

This chapter may suitably be brought to a close by a brief consideration of China's changing world position. It is difficult to anticipate the future influence of the rest of the world on the development of Chinese science and society. However, in the short run her relations with the Soviet Union, America and Japan are clearly major factors. At present most signs suggest, in spite of some to the contrary, continued good relations with the West, Japan and the third world, but with these relations increasingly balanced by improved relations with other socialist countries. This situation and trend should favour a continued strengthening of the legalist formation, including the flourishing of science.

In the longer term, leaving aside the question of possible nuclear war, a major change is likely to come about as Chinese power, wealth and influence in the world, become more consonant with her size and the importance of her culture. At such time, when China will be able to develop more independently, more in keeping with her own specific needs, and more in tune with her own past history, it is probable that this vast and wise civilization will make an appropriately large contribution to social and cultural progress, and not least in the realm of science.

The first period of the People's Republic was one of learning from the Soviet Union; the present period is one of learning from the United States, Japan and the other advanced economies. The next era may well be one in which the rest of the world primarily learns from China, as Europe did to its great enrichment, from the time of Marco Polo well into the eighteenth century.

3 Biomedical Research

The Chinese health care system is unique in several important aspects. Throughout the past three decades, it has been marked by remarkable consistency of policies. To what degree were these guidelines actually implemented during specific periods is, of course, another question. One of the most distinguishing characteristics of the policies is summarized by the dictum 'health care for the people'. This implies, for example, stress on work in rural areas. The training of barefoot doctors, a practice that has spread to other third world countries, is one of the means with which medical treatment becomes accessible in Chinese villages. The practice underlines also the principle of 'health care by the masses', for many of those para-medics are peasants themselves who have returned after training.

China does not regard the health of the people as an isolated phenomenon but rather as one of many conditions which mutually interact. Thus the level of education is another; patients are encouraged to understand the nature of the illness inflicting them and the treatments they have been prescribed (cf. the self-mystification of practitioners in the West). Furthermore, emphasis is laid more on prevention than on treatment.

Successive grass-root campaigns have been conducted through vigorous propaganda to educate the people on ways of preventing specific diseases. These campaigns have largely eradicated endemic and epidemic diseases. As a consequence, while other third world countries still suffer from so-called tropical diseases which are really diseases of poverty, the most common cause of death in China today is cancer, as will be mentioned in the beginning of Part II. That the government may initiate mass-campaigns to combat cancer is an exciting possibility.

The last characteristic of health care in China to be noted is the combination of Western and traditional medicine. This policy may be looked upon as an application of the general philosophy of 'walking on two legs', viz reliance on both foreign and indigenous expertise in the development of the country. In this particular case, the policy is more than appropriate as Chinese traditional medicine is a 'great treasure chest' with a sophisticated theoretical basis and a wide therapeutic armamentarium, on a par with Western medicine. It does not mean that difficulties do not exist, of course. Theoretical concepts in traditional medicine are often vague to the Western (reductionist) mind. As examples, the terms kidney and 'spleen' appearing in an ancient text would refer less to the respective organs than to the corresponding functions, meaning neurophormones and ionic balance regulation, and digestion and transportation respectively. In pharmacology, those Chinese medicines that are herbal in origin often suffer from the problem of standardization in ingredients: plants of the same species but growing at various localities differ in detailed compositions that may or may not affect therapeutic effectiveness. Nevertheless, as discussed in Section 8 and Section 7, tremendous progress has been made in these aspects since 1949; in November 1981, the Chinese Association of Integration of Traditional and Western Medicine was set up. This chapter comprises two parts, which are invited contributions from two authors who are active in the respective fields of work.

Basic Studies*

In 1966 *Science* acclaimed the synthesis of insulin (a protein) in China as a major scientific achievement. Sixteen years later, the Chinese newsweek *Beijing Review* announced that total synthesis of yeast alanine tRNA (a nucleic acid) has been completed.

Chinese investigators commonly demonstrate sound knowledge of Western developments in the biological sciences. They are familiar with current analytical methods and procedures and are skilled in devising useful modifications. Although their number is small, they have been able to produce some excellent results and to train many students. With further support for basic investigations and more contact with Western scientists, they should be able to make rapid progress in the area of basic biological research.

General scientific procedures taught in China are quite similar to those currently employed by Western researchers. Experimental approaches, methods, equipment, and even the format of reporting adopted by Chinese researchers closely resemble those in use in America and Europe.

Chinese laboratories involved in basic biological research in general appear to have adequate equipment and facilities. The government has made every effort to provide modern laboratory equipment for research. Chinese workers have made clever reproductions of Western instruments of the most complicated kind. Factories have been set up for this purpose in Shanghai, Nanjing, Suzhou, Wuhan and Shenyang. In general, the finished products from these factories meet international specifications.

3.1 Protein Chemistry

The synthesis of biologically active insulin in 1965 is one of China's most significant scientific accomplishments. It was performed primarily at the Shanghai Institutes of Biochemistry and of Organic Chemistry. Apart from being a major impetus in the development of biochemical factories, it has led to research on X-ray crystallography of insulin (at the Beijing Institutes of

*By Guo-Fan Hong (Shanghai Institute of Biochemistry, Chinese Academy of Sciences and Medical Research Council Laboratory of Molecular Biology, Cambridge, UK)

Biophysics and Physics and at Beijing University), structure-activity of insulin analogs (at the Beijing Institute of Zoology), and synthesis of other polypeptide hormones. Present work involves the further study of various insulin analogs that includes synthesis, biochemical characterization, X-ray crystallography, and measurements of biological activity and receptor binding.

The approach to enzyme structure and function is both theoretical and experimental. A group at the Shanghai Institute of Biochemistry (the leading biochemical research institute in China) has examined data from the literature relating the modifications of functional groups of a variety of enzymes to their biological activity. From this work equations have been established, from which the number and types of essential groups involved can be calculated. This defines the best experimental conditions for obtaining further results. The investigation covers many hydrolytic enzymes, some of which are now being examined experimentally. Also, it has led to the suggestion that, in many cases, two enzyme substrate intermediates are involved, one of the Michaelis type and the other an acyl enzyme intermediate in the case of proteinases esterases. This type of mechanism has been generalized to other systems and shown to be compatible with data from literature.

The study of some proteolytic enzymes has been approached experimentally, such as the action of sulphite on the S-S bridges of trypsin in the presence of p-chloromercuribenzoate, and the study of p-nitrophenacyl bromide as an inhibitor of insulin, ribonuclease, papain and α-chymotrypsin.

Further investigation on papain (photo-oxidation of histidine and of tryptophan residues as well as the oxidation of the latter by N-bromosuccinimide at pH 5.0) has established that one of each of these residues was essential for papain. It has also been shown that a polymer of leucine could be obtained by incubating leucineamide with papain at pH 7, as well as with a spleen preparation apparently different from known cathepsins. Other observations have led the scientists to propose an original spectrophotometric method for determining protein SH groups by the absorption at 420 μm of the complex formed between these groups and 1,4-naphthoquinone.

Four strains of tobacco mosaic virus (TMV) have been compared by serological methods, electron microscopy, fingerprinting of the proteins, etc. The results suggest a close genetic parentship between the various strains (identical terminal threonine) but a closer similarity between the YMV and DDV strains. During these investigations, a method using chloroethanol to precipitate the RNA was suggested for obtaining pure virus proteins. The proteins and nucleic acids can be recombined to form active virus. The amino-acid composition and the C-terminal pentapeptide structure of YMV suggest that it is a new strain of TMV.

Work on the molecular properties of muscle proteins is commencing at the Shanghai Institute of Biochemistry. At the same time, the physiological study of muscle is being carried out at the Institute of Physiology. Scientists at the former Institute are investigating, with the electron microscope, a number of para-crystalline forms of tropomyosin and paramyosin from a variety of biological species. The two proteins, on the basis of electron microscopic periodicities, are quite different. However, when present in the same muscle, as in some molluscs, a close genetic relationship may exist between them as inferred from the intraperiod fine spacings of 140 Å found for Anodonta tropomyosin. Such fine spacings are characteristic of paramyosin from the same origin; the usual spacing is 200 Å for other tropomyosins. Tropomyosins of various origins were also examined immunologically and by fingerprinting, and some interesting evolutionary aspects of shark and ray proteins were obtained.

3.2 Nucleic Acid Chemistry

The transfer ribonucleic acid (tRNA), a key molecule in the translation of the genetic code into specific protein structures, has been directly synthesized for the first time by a Chinese team. The construction of the tRNA replica, which has taken years of effort by a large number of workers, was crowned by a laboratory test showing that the replica is capable of biological activity.

The synthesis of yeast alanine tRNA is a product of collaborated efforts from several institutes of the Academy of Sciences (including the Institutes of Biochemistry, Cell Biology and Organic Chemistry in Shanghai and the Institute of Biophysics in Beijing), the Department of Biology of Beijing University and the Shanghai Number 2 Reagent Factory.

As with insulin synthesis, the success of yeast alanine tRNA synthesis will promote the production of reagents and enzymes for nucleic acid research work, as well as the production of a number of ribonucleic acid derivatives which might be of medical use. Furthermore, the large amounts of comparatively rare ingredients that have been assembled should make it feasible to undertake a systematic investigation of how changes in the chemical structure of a tRNA molecule affect its biological activity. The Chinese will be looking with particular interest at the roles played by the modified nucleotides in the functioning of tRNA.

3.3 Enzymology

Investigators at the Shanghai Institute of Biochemistry have partially purified succinic dehydrogenase and established, independently of other workers, that succinic dehydrogenase was a metalloflavo-protein containing non-haematin iron. It was demonstrated that succinic dehydrogenase was linked to the cytochrome c part of the chain through a common velocity-limiting factor with NADH dehydrogenase. These investigations led to an examination of cytochrome C from mammalian heart muscle and from yeast. The flavine prosthetic group of purified succinic dehydrogenase was separated after trypsin and chymotrypsin digestion from the proteins, and fractionated by electrophoresis into four flavine derivatives containing various amounts of attached amino-acids. These are inactive in the D-amino-acid oxidase test and are therefore distinct from FAD. These flavin-adenine peptides do not show changes of absorption maxima in alkaline or acid solutions like FMN and FAD. This is probably due to the peptide linked to the isoalloxazine ring. The nature of the peptide and type of bond to the isoalloxazine ring awaits further characterization. Continuing this work on the respiratory chains, they recently compared the action of various inhibitors on choline and L-α-glycerophosphate oxidase, and concluded that both systems contain an antimycin A or BAL sensitive step, whereas cytochrome b is concerned only with choline oxidase.

Further work on the relationship between the succinic and NADH oxidase systems gave no experimental support for the suggestion that endogenous cytochrome c was directly linked to the succinic oxidase-NADH oxidase chain through a phospholipid. The oxidation of succinate in function of the concentrations of 2,4-dinitrophenol has recently been examined. At low concentrations of DNP respiration of succinate is increased, while at higher concentrations, after a short stimulation, there is an inhibition which can be prevented by adenosine triphosphate (ATP), which also prevents the inhibition of respiration that is obtained with amytal or arsenate. On account of this observation the Chinese postulate that some activation energy is required for succinate oxidation.

Recent research shows that no inhibition of respiration by DNP is observed when succinate is oxidized with 2,6-dichlorophenolindophenol – showing that DNP does not inhibit at the dehydrogenase level. In addition to ATP, restoration of respiratory activity can also be obtained in the presence of substrates of the NADH-linked dehydrogenases. The Chinese have suggested that this high energy intermediate may be formed by the reversed reaction of the succinate linked endogenic reduction of NADH in the presence of DNP. This reduction is inhibited by amital, dicoumarol and DNP and is completely dependent on magnesium, manganese and probably another metal which can also be removed by EDTA. Based on a comparative study of the competitive inhibition by aged haematin on various flavoprotein enzymes, it is now agreed that the inhibitory effect of this compound depends on the acceptor system. The similarity of response on succinate and NADH oxidation (heart muscle mitochondria) with that of choline (rat liver mitochondria) and of α-L-glycerophosphate (rabbit skeletal mitochondria) has led to the suggestion that the two last-mentioned are also flavoprotein enzymes.

Whole-body irradiation was found to induce mitochondrial swelling within 30 min to 3 hours. This effect was greater in the presence of Ca^{2+} or thyroxine, and the conclusion is that structural damage to the mitochondria has been induced. However, respiration and oxidative phosphorylation, which had been somewhat inhibited, became normal again after four hours.

The oxidation of NADP in rat liver microsomes and supernatant was examined, and the use of various inhibitors showed that cytochrome oxidase is not involved, but that elements of both cell fractions seem to be necessary for full activity.

Some investigators are working on photosynthesis. They have devised kinetic equations for photosynthesis based on the assumption of two limiting reactions when flashing light is used – one photochemical and one biochemical. The relation between flash yield and flash period predicted by this equation agrees with experimental results recorded in the literature. The equation can be simplified under certain conditions and permits the calculation of the rate constant for both the limiting reactions as well as the concentration of the photosynthetic unit. Electron spin resonance experiments are also being performed on chloroplasts.

There is also practical interest in enzymes for use in the chemical, pharmaceutical and agricultural industries. The basic biochemistry has come mostly from the Institute of Biochemistry, with some papers from Beijing Medical College, the Institute of Microbiology, and Zhongshan, Beijing, and Nankai Universities. In addition, work has been reported on the characterization of oxidative enzymes at the Institute of Biochemistry and of rabbit muscle glyceraldehyde-3-phosphatase at the Institute of Biophysics.

3.4 Virology and Cancer Research

The development and characterization of vaccines are actively pursued at many institutes in different cities in China. There is also much interest in the relation of viruses to cancer.

Vaccines are prepared using attenuated live measles virus, inactivated type-B encephalitis virus, attenuated live influenza virus, absorbed epidemic cerebrospinal meningitis agents, asthma and attenuated live poliomyelitis virus, as well as vaccines against smallpox. An inactivated vaccine has been prepared against type-B Japanese encephalitis virus grown in cell culture and used with good results. Attempts to develop a vaccine against hepatitis B virus and to improve the Sabin polio vaccine are under way.

Diagnostic virology is performed using cell culture, and simple comparisons of different viruses in cell culture are performed. Fluorescent and electron microscopies are used. Articles on antiviral chemotherapy against influenza have been published, and the use of polynucleotides as interferon inducers has also been reported. Rickettsia, especially the trachoma agent that was isolated in China in 1957, are also studied.

The Chinese institutes have facilities for cell culture, although some of the media are still imported. Cell culture is used in virology and cancer research. Chinese biologists have established a human diploid cell strain, and have grown human oesophageal cancer epithelial cells as well as liver cancer cells in vitro. They have obtained valid results that should be of general use.

Chromosome-banding techniques have been applied at the Hunan College of Medicine, Institute of Oncology, Beijing Medical College and Institute of Genetics. The techniques are being applied to clinical material and use in experimental oncology.

Work on radiation cytogenetics is under way at Suzhou Medical College. The immediate objective is to repeat basic results obtained elsewhere. The Institute of Experimental Biology is doing the work on chromosome structure.

Scientists at the Institute of Zoology have been studying transplantation of nuclei from somatic cells into eggs, between different species and even between subfamilies of fish, especially goldfish and betterlings. Their results have extended earlier Western work and are generally known in China. Some work on reconstituting membrane adenosine triphosphatase in cancer cells is carried out at the Institute of Zoology.

There is an abundance of cancer research in China. The major efforts are clinical and epidemiological in nature, with relatively little laboratory or experimental work (next section). Cytogenetics, cell fusion, and cell culture have been used. The Chinese are attempting to promote immunogenicity by cell fusion, nuclear transplantation, and nucleic acid injection. They are also interested in studying possible virus involvement in certain human cancer. Chemical carcinogenesis is little studied, although the possible importance of chemicals in the aetiology of human cancer is recognized. In some cases attempts had been made to isolate or remove chemical carcinogens from the home and work place.

Attempts to isolate new anticancer compounds are being made, especially at the Shanghai Institute of Materia Medica. Microbiological systems are often used in initial screening. Some new drugs have been isolated.

There is interest in reproductive physiology. A few papers on prostaglandins have appeared. The limited research efforts were firmly practical, akin to those of state agricultural experiment stations in the United States. Keshan disease (the endemic myocardial disease of unknown aetiology) is studied at Harbin and Jilin Medical Schools. Modern techniques are used in the search for new drugs, especially from traditional medicines (see section 7 below).

3.5   Principal Research Centres in Basic Studies

ACADEMY OF SCIENCES
Inst. of Biochemistry (Beijing)
The Institute of Biochemistry has been the leading research institute in the synthesis of biologically active insulin and tRNA. These two works led to an increase in China's capability to synthesize peptides and oligonucleotides, a capability now exploited for synthesizing modified insulins for structure-activity studies and several other polypeptide hormones including oxytocin, vasopressin, angiotensin II, and glucagon, which are used in medical research.

The institute was found in 1950 as a joint institute of Physiology and Biochemistry, but it was split up in 1958. It has about 500 personnel, including about 250 research workers and 90 staff members in a nearby biochemistry factory. Areas of research interest included protein synthesis (methods, modified insulins, polypeptide hormones), nucleic acid synthesis, structure of nucleic acids (tRNAs), enzymology, liver cancer (diagnosis, α-fetoprotein, hepatitis B virus, therapy), and plant viruses. There is also an instrument division.

Several projects involve applied research. The Biochemistry Factory and Instrument Division prepares reagents and makes equipment.
Inst. of Biophysics (Beijing)
This Institute has been involved in X-ray diffraction studies of native and modified insulins. The Institute was founded in 1958 as an offshoot of the

older Institute of Physiology and Biochemistry in Shanghai mentioned above. The Institute does not have one central laboratory building, but is scattered among various other institutes. There are five divisions; Radiation Biology, Molecular Biology, Sensory Receptors, Submicroscopic Structure of Cells and Technical Projects and Instrumentation.

Inst. of Cell Biology (Shanghai)
The physiology of reproduction and cancer research are among its chief topics of interest.

Inst. of Zoology (Beijing)
Major projects involve the reproductive endocrinology and the nuclear cytoplasmic relationships. The nuclear transplantations done here are generally of good quality, and some of them are unique. See Section 2 of the following chapter for more information.

Inst. of Physiology (Shanghai)
There are five divisions: acupuncture anesthesia, muscle biophysics and nerve-muscle trophic relationships, sense organs, hypoxia, and reproduction.

Inst. of Genetics (Beijing)
This institute is mainly concerned with plant breeding, transplantation of fertilized sheep eggs, and microbial genetics. Most of the work is directed towards practical applications; some will be examined in Sections 1 and 2 of the next chapter. Established in 1951 as an office, it was enlarged to be of institute status in 1959 and moved to its present building in 1964.

CHINA ACADEMY OF MEDICAL SCIENCE
Inst. of Cancer Research (Beijing)
This institute is concerned with human cancer primarily from the viewpoints of treatment and epidemiology. The epidemiology, primarily of esophageal cancer, was of high quality and great interest. The laboratory studies involved testing suspected foodstuffs for carcinogenicity.

Clinical Practice*

Some general and illustrative examples of Chinese medical practice will be discussed in this second part of the chapter. With so broad a scope to be covered, attention is mainly restricted to achievements in recent years. It is well known that in China there are two kinds of medicine, each of which has its own theoretical basis and therapeutic methods. Equal emphases will be given here to the two, namely Western and traditional medicine.

3.6 Cancer

Since 1949 the pattern of disease in the country has changed entirely. Some acute epidemics and communicable diseases have been wiped out or largely eliminated, while cancer has become the major killer. For this reason the

*By Neng-Ren Jin (Institute of Haematology, Beijing Medical College, Beijing)

government has paid increasing attention to cancer research. Institutes for such research were established in the late 1950s and early 1960s. According to the policy of public health, mass-survey and systematic screening became the key techniques in the control and prevention of many forms of cancers.

In the 1970s a nationwide mass-survey began with the goal of determining the incidence of the common types of cancer. This investigation, organized by the Chinese Academy of Medical Science (CAMS), was started first as a pilot project in Lin-xian County of Honan province, where a high incidence of oesophageal cancer had been evident. The work was subsequently extended throughout the country. This large-scale research programme involves eight research institutes, 1.8 million barefoot doctors and many special coordinating groups established to deal with cancers. Most of the provinces, municipalities and autonomous regions have been mapped to show the relative incidence of each type of cancer, but the programme is still going on. Broadly speaking, the cancer of highest incidence by region is as follows: oesophageal cancer in the north, liver cancer in the east, and naso-pharyngeal cancer in the south. Lin-xian County, inhabited by some 110 000 people, has an annual incidence of oesophageal cancer as high as 1 to 1.5 per 1 000, which is about 50 times greater than the average world rate. In Qi-dong County of the eastern province of Jiangsu, the annual incidence rate of liver cancer is around 0.53 per 1 000.

Based on results produced so far from the mass-survey, remarkable progress has been made in detecting and treating cancers in their early stages. It used to be thought that early diagnosis of liver and oesophageal cancers was particularly difficult, because they show no symptoms in their early stages, and when the symptoms do occur it used to be too late to save the patient's life. As a result of recent work, however, the case is no longer true. The method used widely in China for the detection of liver cancer is α-foetoprotein (AFP) assay. AFP is a substance produced in the liver of a newborn, its production stopping when the baby is about 3 months old: hence the designation 'foetal'. On the other hand, for some reason it is secreted by a malignant liver and is therefore considered to be an indicator of liver cancer. AFP assay is accurate, requires only one drop of blood from the individual's ear or finger, is simple enough for barefoot doctors to handle, and thus finds ready application in large-scale surveys. Furthermore, a radioimmunoassy technique has now been developed that gives a sensitivity of 10 nanograms AFP per millilitre of serum. In 1977 a survey of 637 000 people, conducted by the Shanghai First Medical College, turned up 213 cases of liver cancer, 76 per cent of them in early treatable stages. Partial surgical removals of livers were subsequently performed, enabling 58 per cent of the patients to survive at least one year, 32 per cent three years, and (as of 1983) 27 per cent five years or more. For oesophageal cancer, the diagnostic test consists of the screening of oesophageal cells, which are collected by either a single-lumen tube or a double-lumen tube with an abrasive ballon. These collection techniques, originated from the Chinese, have made detecting oesophageal cancer in its early stages possible and its treatment more effective.

The Chinese are equally advanced in the area of modern physical methods. The case of liquid crystals serves as a good illustration. Diagnosis of tumours in breasts and skins with the help of liquid crystal thermography is routine at the Tianjin Institute of Tumours, Shanghai First Medical College and Hubei Medical College. The liquid crystal screens in use at the Hubei Medical College have been designed in cooperation with Wuhan University, and consist of micro-encapsulated cholesteric films. The diameter of each individual capsule is 10–30 microns. The temperature range in which these films change colour is 3–4°C, with a sensitivity of 0.5°C. Incidentally, general research on medical applications of liquid crystal technology is quite active in the country. In Shanghai, for example Fudan University, East China Institute of Chemical Technology, Shanghai Institute of Dyes and Pigments and Huashan Hospital are collaborating in the development of a liquid crystal viewer for three-dimensional X-ray examination of human bodies. In the Enshi Municipal Hospital of Hubei province a liquid crystal display for X-ray images is being studied. The goal is to replace the current X-ray luminescence screens which are by comparison expensive, poor in contrast and low in absolute brightness. Also in Fudan University, there is a project directed towards the same goal that instead involves rare earths as sensitizers (see p136).

Regional epidemics of cancer are a focus of attention in the mass-survey. In Lin-xian County the high incidence of human oesophageal cancer is paralleled by an excess of gullet cancer in domestic chickens. Since animals have a narrower environment than human beings and shorter latent periods in developing cancers, they provide an excellent means of identifying potential carcinogens. Through countless experimental investigations, the CAMS in conjunction with the Honan coordinating research group, finally identified the main etiological suspect as a pickled vegetable mix. A Lin-xian speciality, the mix is sometimes fed to chickens along with food scraps. Besides pickled vegetables, fungal contaminants, nitrosamines and personal eating habits are also considered as contributing factors. The Liver Cancer Research Institute (Qi-dong County) with help from CAMS has pinpointed Aspergillus oryzae, a type of fungus that grows on corn and peanuts, as a specific carcinogenic agent for liver cancer. It is also agreed that hepatitis B is a predisposing factor for liver cancer. Stomach cancer, a prevalent kind of malignant tumour in China, is studied at the Beijing Institute of Cancer Research. Although it is well known that the intakes of mouldy or smoked food and alcohol may lead to stomach cancer, the researchers now believe that 'atrophic gastritis', a stomach ailment which results in the inflammation of the stomach wall, is closely related to the disease. It was found that 30 per cent of the population in areas with a high incidence of stomach cancer suffer from various chronic stomach diseases, among which atrophic gastritis is the most predominant.

China has taken significant strides in cancer therapy. Methods of treatment include surgical operation, chemotherapy as well as radiation therapy. Nuclear medicine is a fast developing branch of Chinese medical science. Research in this field began in the 1960s, and now in addition to cancer centres more than 700 hospitals make use of radioisotopes to diagnose and treat diseases. (Refer to Chapter 7.4 concerning the national supply of radioisotopes.) Radiation therapy is widely applied in treating chronic leukaemia, lymphoma, as well as cancers of brain, kidney and uterus. In 1980 the Chinese Society of Nuclear Medicine was set up and it publishes a journal entitled *Chinese Nuclear Medicine*.

The Capital Hospital in Beijing has developed a new therapeutic method that dramatically reduces the mortality rate of choriocarcinoma, a highly malignant cancer in gynaecology. In contrast with the traditional treatment – removal of the womb, followed by radiation therapy – the research group in the hospital treats the metastatic choriocarcinoma with high doses of chemotherapy, and results obtained have been extremely encouraging. The chemotherapeutic agents used are 6-Mercaptopurine (6MP), 5-Fluorouracil (5-FU), Actinomycin K, Methotrexate (MTX) and AT-581. The choice of medicine is individualized according to the conditions in each case, in particular the sites of the metastases. Treatment is continued until attainment of complete recovery, the criteria for which are absence of all clinical symptoms, complete disappearance of pulmonary metastatic shadows and negative male toad test by the urine concentration method. After this, one to two courses of therapy are administrated to ensure complete cure before discharge. A follow-up survey of 618 cases shows that by 1980 no patient had died within five years after treatment. Furthermore, of the 159 young women treated with chemotherapy without removing their uteri, 85 per cent gave birth after discharge. Their babies showed no differences from those of normal mothers either in physical or mental development. Based on laboratory study and clinical practice, the research group has worked out a new theory on the origin and development of the disease. According to this theory, the tumour cells spread from the uterus to other parts of the body with the return flow of blood through the vein. This model is now being investigated by foreign cancer specialists.

Acute leukaemia had very poor prognosis 20 years ago. Thanks to the applications of new chemotherapeutic agents, therapeutic regimes and effective supportive care, remarkable progress has been achieved in leukaemia treatment. The Institute of Haematology at the Beijing Medical College reported a series of cases of acute non-lymphocytic leukaemia, which were treated with Hydroxyurea (HU), Cytosin Arabinoside (Ara-C), Harringtonine, Daunomycin and prednisone. The complete remission rate is up to 80 per cent.

Apart from chemical drugs, Chinese herbs and medicines are also widely and effectively used in combating cancers. For example, a traditional medicine called 'Sanpin' has been successfully used in the treatment of cervical cancer in its early stages. This process involves the insertion into the cervix of a 'Sanpin' suppository, which causes the cancerous tissues to coagulate and fall off. Simpler, safer and less expensive than surgery and radiation therapy, this operation can be carried out at even a small commune hospital in rural areas. Jiangxi Maternity Hospital has reported its investigation of 170 patients suffering from the early stages of cervical cancer. After being treated with 'Sanpin', 78 patients have had no recurrence of the disease so far (after five years). One patient gave birth four years after treatment.

3.7 Surgery

Surgery is a most active field of medical science in the country. Surgical practice in China covers the whole gamut of what goes on in Western countries. Almost all rare and difficult operations in open heart surgery, neurosurgery and microsurgery are carried out. These branches of surgery emerged in the 1950s and 1960s and have since been rapidly developing.

Fuwai Hospital and Cardiovascular Institute, one of the biggest cardiovascular centres in China, opened in 1958 as a chest hospital, under jurisdiction of CAMS. Possessing seven research divisions and 400 beds, it treats patients with hypertension, coronary artery disease, valvular heart diseases and congenital heart diseases. Much attention has been paid to surgical practice. Approximately 150 operations of open heart and vascular surgery are performed annually including valve replacement, prosthesis, coronary artery bypass surgery and neonatal cardiovascular surgery. The mortality rate in open heart surgery has been reduced significantly by a method of selective hypothermia combined with extracorporal circulation.

Neurosurgery has seen outstanding advances in the last two decades. There are now over 300 hospitals through-out the country performing neural operations. Beijing Hsuanwu Hospital and Neurosurgery Institute, established in 1959, have made great contributions to the treatment of brain injury and brain tumours. The integration of research work and clinical practice has resulted in both lower mortality rates in brain tumour surgery and better preservation of normal nerve functions. For example, the rate of removal of total acoustic tumours has reached 84 per cent; at the same time the probability of facial nerve preservation has risen to 90 per cent and mortality has dropped to less than 3 per cent. For pituitary tumour, intracranical and trans-sphenoidal approaches are now adopted under microsurgery. The tumour can thus be removed completely without interference with the normal endocrine function of the pituitary.

Chinese surgeons are strong in the field of severe burns. In 1958, Guangci Hospital (today the Ruijin Hospital) attached to the Shanghai Second Medical College saved the life of a worker with burns on 90 per cent of his body surface (23 per cent among which were third degree burns). Since then departments and institutes for severe burns were set up in large hospitals and big cities. Research on severe burns is now scattered among a number of centres. Many patients, some with even larger percentages of burned surface and in worse conditions than the patient mentioned above, have recovered satisfactorily. The procedure established by Chinese surgeons is highly successful and has aroused much interest among colleagues abroad. Its principle consists of the early excision of scar zones after a few days, followed by application of perforated pigskin graft, with orifices of the perforations covered by autografts taken from the surface of the patient's scalp, which is often undamaged. Since they are extremely thin, such autografts may be taken repeatedly at the same location every five days. The cure rate for patients with severe burns is now about 93 per cent overall. The rate for patients with burns over 80 per cent of skin surface is 40 per cent, and that for those with burns over 90 per cent of skin surface (70 per cent being third degree burns) is 15 per cent. These figures represent a remarkable success that brings China to a leading position in this field.

The development of burn treatment has pushed plastic surgery to higher levels. As major burns heal, victims require further treatment by plastic surgeons to reduce deformity and restore functions. Since the patients have very little normal skin left, large homografts inlaid with small autografts from the patients' own scalps have also been applied to repair large wounds after late-stage surgery to remove contractions. Thus the patients can move their limbs and fingers more freely and smoothly. Some of them even have taken on simple jobs.

Microsurgery emerged in the early 1960s. The Shanghai Sixth People's Hospital, the pioneer in microsurgery, successfully replanted a completely severed hand in 1963 and then in 1966 replanted a severed finger. In the early period, low-power optical microscopes were used in microsurgery, and success rates attained were 83 per cent for cases involving the limbs and 58 per cent for those involving fingers. Since 1973 two-man surgical microscopes, and atraumatic (non-injury) needles of 18 microns diameter, have been introduced into microsurgery by Shanghai Sixth People's Hospital. These advances raise the success rate to 92 per cent.

The application of new microsurgical techniques has opened a new era in microsurgery. The Shanghai Sixth Hospital reconstructed an amputated thumb by grafting on the patient's second toe. It is well known that removal of the second toe does not seriously harm the foot's functions but, in contrast, to lose the thumb is to lose half of the hand's functions. Another example was free fibula transplant with blood vessels attached. The Shanghai Sixth People's Hospital performed China's first such operation on an eight-year-old boy, whose humerus has been eaten away by osteomyelitis. Because of the immediate establishment of blood circulation on the bone graft, bone withering and necrosis were prevented. Thus, with strong resistance against infection, the graft survived and the boy regained the function of his arm quickly. Later, the specialists in the Shanghai Ninth People's Hospital designed and installed a new hand for a patient who lost both hands in an accident. The hand, shaped like a pincer, consists of a metal metacarpal bone and two of the patient's own toes. It can manage most of the work done by a normal hand.

The repair of a damaged oesophagus has been a serious problem to be tackled by surgeons. The Shanghai Sixth People's Hospital successfully reconstructed an oesophagus damaged by chemical burns with free transplant of a segment of the patient's intestines. This kind of operation can now be performed in other big hospitals. Another advanced technique is the free transplant of the greater omentum, a fold of the peritonium. This technique has been applied to repair leg ulcers, scalp ruptures, cranial damage and semi-facial atrophy. The Beijing Jishuitan Hospital, Shanghai Ninth People's Hospital and the General Hospital of the People's Liberation Army have all treated successful cases.

Recently the Affiliated Hospital of Bangbu Medical College, Anhui province, reported treating eight patients with secondary obstruction lymphoedema of upper and lower extremities with micro-lymphangiovenous anastomosis. In all cases the immediate postoperative and follow-up results are very satisfactory. All in all, China is at the forefront in microsurgery. But it is still a long way to go both in widening the scope of microsurgery and in raising its technical levels.

Organ transplantation in China initiated in the 1960s. Though the gap in expertise widened between China and some Western countries during the Cultural Revolution, this field has developed rapidly in recent years. Until 1981, when the Symposium on Organ Transplant was held in Wuhan, Chinese surgeons and haematologists had performed a number of kidney transplants and some liver, lung, and bone marrow transplants.

According to figures released at the First National Congress on Urology convened by the Chinese Medical Association in 1981, the total number of kidney transplants then already exceeded 900. The first successful operation was performed at Zhongshan Medical College, Guangzhou. The Beijing Friendship Hospital has completed over 120 kidney transplants. The Zhongshan Hospital, which is attached to Shanghai First Medical College, reported the case of a patient surviving over six years after the operation. In total, the one-year survival rates of the kidney graft and the patients are on average 50 per cent and 75 per cent, whereas the two-year survival rates are 43 per cent and 47 per cent, respectively. These figures are very similar to those from the International Register Office.

The Institute of Organ Transplant, Wuhan Medical College and Ruijin Hospital, Shanghai Second Medical College are intensifying their efforts on liver transplant. The half-year survival rate remains low (11 per cent): much needs to be done in the future to improve the technique. Those for lung, heart and pancreas transplants are in a preliminary stage.

Bone marrow transplantation is an active field in haematology. The Institute of Haematology (formerly the Department of Medicine, People's Hospital) in the Beijing Medical College performed, in 1964, China's first syngeneic bone marrow transplant on a patient suffering from severe aplastic anemia, using bone marrow cells from her identical twin. The patient completely recovered and her marrow function was fully restored after this procedure. In 1981 the Institute reported China's first successful allogeneic bone marrow transplant, which was performed on a 20-year old woman suffering from acute monocytic leukaemia. She is now leading a nearly normal life. Six months after the transplant, examination of the patient revealed that her chromosome, blood group, and iso-enzymes were identical with those of the donor (her brother) but different from those she had had previously. Thus bone marrow transplants realize the possibility of curing acute leukaemia, severe aplastic anaemia, acute radiation disease and some other refractory diseases.

3.8 Pharmacology

China's pharmaceutical industry has developed at a fast pace. During the 1950s the country became independent of important drugs as well as chemicals needed for drug manufacture. By 1958 she was able to export sulfonamides, penicillin and vitamins. Now nearly 1 000 kinds of medicine and more than 3 000 preparations are produced. In addition, some 3 000 traditional medicines are manufactured. The products not only meet domestic needs but also benefit people abroad.

Advanced medical research, including research and development activities in pharmaceuticals, is conducted within the Chinese Academy of Medical Science. In addition, CAMS coordinates research programmes throughout the country. The most advanced pharmaceutical research takes place at the Institute of Materia Medica in Beijing, Shanghai Institute of Materia Medica, Beijing Institute of Microbiology, Shanghai Institute of Biochemistry (the last three being under the administration of Chinese Academy of Science) and several major institutes under provincial control or the dual jurisdiction of the Public Health Ministry and provincial authorities. Current emphases centre mainly upon cancer, cardiovascular disease and contraceptives. The Chinese Academy of Traditional Medicine also carries out research on pharmacology, in relation to Chinese herbal, animal parts-derived and mineral medicine.

The Shanghai Institute of Materia Medica has five research divisions. Its chief concerns are the studies of new drugs and antitumour agents, treatments of chronic bronchitis and cardiovascular diseases, as well as basic research in support of applied projects in drug synthesis or drug metabolism. Since 1958 over 2 000 substances have been screened for antineoplastic effect. The test procedure is that, after undergoing preliminary testing for acute toxicity, the experimental compound under study is used to inhibit the development of transplantable tumours in rodents. The Institute have isolated Vincristin and Vinblastine from the plant *Vinca rosea*, and synthesized a series of derivatives of 6-mercaptopurines, an antimetabolites of purines.

The isolation of alkaloids from San-Qien-Shen is considered a great achievement of the Institute. San-Qien-Shen, or Cephalotaxus fortunei Hook f., is an evergreen tree extensively distributed in the south of China. In the past it has been the source of a folk medicine for treating tumours. In the 1970s the San-Qien-Shen Research Cooperative Group of the Institute isolated and identified more than ten types of alkaloids from the branch, leaves and bark of the tree. Among them, two natural ester compounds, namely San-Chien-Shen ester alkaloid I and high San-Qien-Shen ester alkaloid II, were proved to be identical to harringtonine I and homoharringtonine II. Furthermore, semisynthetic harringtonine is now manufactured. All these drugs have been proved, on the basis of pharmacological analysis and animal traits, to be extremely effective against malignant diseases, particularly against brain tumours and acute leukaemia. A number of reports show that the complete remission rate of acute non-lymphocytic leukaemia in adults, treated by harringtonine in combination with other drugs is 60 to 80 per cent.

The Institute of Traditional Chinese Pharmacology is affliated to the Chinese Academy of Traditional Medicine. Two of the major tasks of the Institute are to identify effective traditional drugs and prescriptions for treating and preventing common diseases, and to develop new drugs. After it became known that the parasites for malignant malaria (*Plasmodium falciparum*) had developed resistance to chloroquine, researchers at the Institute started to explore for new drugs from Chinese herbs. They studied all antimalarial prescriptions mentioned in ancient medical literature and noticed that the herb Sweet wormwood was often mentioned in prescriptions. Through a large number of experiments the researchers finally extracted an effective monomer from Sweet wormwood, and named it 'Qinghaosu'. With a chemical structure entirely different from those of other antimalarial drugs based on Quinine, it was, nevertheless, proved effective in mice infected with malaria. Similar encouraging results were obtained with monkeys. Since 1975 'Qinghaosu' has been used in ten provinces, municipalities and autonomous regions. A report on 6 000 cases has been released by a joint research group organized by the Chinese Academy of Traditional Medicine and the Chinese Academy of Science. According to this report 'Qinghaosu' is superior to Chloroquine and other drugs, giving quicker results and lower toxicity, and is effective for all types of malaria, including the dangerous cases of cerebral malaria and malignant malaria that resist chloroquine.

Long experience in the administration of traditional medicine shows that they are highly effective and that most kinds of drugs have no toxicity and little or no side effects. The varieties of traditional medicine used in clinical practice have increased. About 90 per cent are plants (roots, stems, leaves, flowers, fruits, seeds, barks and peels), the rest animal substances and minerals. There are also some 2 000 types of herbal medical preparations of definite standard compositions. The tasks before Chinese pharmacologists are, first, to study the basic theory of Chinese herbal medicine, second, to identify the effective components in the traditional drugs from ancient prescriptions, and third, to synthesize new drugs and to modernize the processing of traditional drugs. The production methods for traditional Chinese medicine are constantly being improved by the introduction of modern technology. In place of decoctions, some herbal medicines are now manufactured in the form of sugar-coated tablets, capsules, injections or instant beverages, which can be taken right away after mixing with wine or boiled water.

3.9 Chinese Traditional Medicine

Chinese traditional medicine, one of the world's oldest medical sciences, dates back to 1 000BC. In the first half of this century traditional medicine officially fell into disrepute but, after 1949, it regained government support and during the last 30 years has been widely used.

Since 1949 many hospitals for traditional medicine have been set up in almost every province and big city. As a rule they have departments of internal medicine, surgery, dermatology and acupuncture, and conduct research on topics of individual choice. For example, chronic aplastic anaemia is a focus of attention at the Xiyuan Hospital of the Chinese Academy of Traditional Medicine. Hepatitis and nephritis are the main topics of research at the Beijing Traditional Medical Hospital. In addition, there are also departments of traditional medicine within general hospitals. In some hospitals special wards have been set up for the purpose of investigating advantages that may be gained from the integration of traditional and Western medicine.

The Department of Traditional Medicine in the Shanghai Sixth People's Hospital is eminent for its outstanding study on theoretical traditional medicine. It has put forward thorough explanations and advanced some new interpretations of the classical theories of 'Five Elements' and 'Yin-Yang', and of certain ancient medical texts such as the *Yellow Emperor's Internal Medicine*, a book from the Warring States Period (473–211BC), and *On Febrile Diseases*, by Zhang Zhongjing of the Han Dynasty.

The system for education in traditional medicine has taken an entirely new look. In the past, the training of traditional doctors was confined to the same family from one generation to the next (grandfather to father, father to son). Now, in addition to such 'invisible' schools, over 20 traditional medical colleges have been established since 1950. The graduates from these colleges now constitute an important part of the total of 260 000 traditional doctors, although the veteran and experienced traditional doctors are still playing a vital role in the medical service.

Recently the first step of computerization was taken in the field of traditional medicine. Some of the rich experience codified by individual veteran doctors of traditional medicine has been stored into computer data base. The Beijing Traditional Medical Hospital cooperating with the Automation Research Institute of the Academy of Science, and the Shanghai Sixth People's Hospital with the Shanghai Automation Research Institute have independently worked out the computer software (cf p191). Now, with the help of the computer which stored the experience of veteran doctors, nurses can diagnose patients accurately even when the doctors are absent.

Acupuncture is an important component of Chinese traditional medicine. It consists of the insertion of fine needles into one or many 'acupuncture points' on the body's various 'meridians'. Each needle is pushed deeper until the feeling of 'deqi' (tingling, distension, heaviness, numbing) is experienced by the patient. The needle point is stimulated by manual rotation, electrical impulse or moxibustion. According to ancient Chinese medical theory, acupuncture cures specific diseases by adjusting the relationship between 'Yin' (materials) and 'Yang' (functions), promoting communication between the channels and collaterals, and regulating the vital energy and blood flows. In other words, acupuncture can harmonize the metabolism of the body and increase its resistance to diseases.

Clinically, acupuncture has been used to treat about 300 different diseases with good anti-inflammatory and analgesic effects. The Chinese Academy of Medical Science has investigated 600 cases of acupuncture therapy of coronary heart disease. The results show that acupuncture is effective in relieving symptoms and in eliminating angina pectoris. Electrocardiograms taken before and after the therapy have revealed an effectiveness rate of 53 per cent, and in some cases ultrasonic cardiography, cerebral hemadromography and measurements of the cardiac output have confirmed that acupuncture can improve coronary circulation, cerebral circulation and left-heart function. Many other reports indicate that acupuncture is capable of certain therapeutic effects in various types of paralysis as well as in some eye, ear and nose diseases.

The regular use of acupuncture as a pain deadener for surgical procedures was initiated in the early 1960s by some medical institutes. Before long the use of acupuncture as an analgesia or hypalgesic anesthetic spread rapidly. By 1977 more than two million operations had been performed under acupuncture anesthesia. Its effect is comparatively stable in 20–30 kinds of ordinary operations and, generally speaking, it is particularly effective in head, neck, and chest surgery. It has also been widely adopted in thyroid, abdominal tubal ligation and caesarean section operations as well as the partial removal of stomach and the removal of prostate or lobes of the lung.

The Beijing Municipal Tuberculosis Research Institute has delivered a report on 1 200 lung operations performed under single acupuncture anaesthesia and stated that the success rate reaches 96 per cent. Through experiment and practice the Institute has reduced the acupuncture points needed from the original 40 points that spread throughout the limbs, to just the 'San-Yang-lo' and the 'Hsi-men' points. Both located under the ear, the former point is strongly anaesthetic and the latter, sedative. Another investigation of 1 000 cases of caesarean section operations performed under acupuncture anaesthesia, undertaken by the Beijing Obsterics and Gynaecology Hospital, shows that the effective rate is 76 per cent.

Why can acupuncture be used in surgery? What is the mechanism of acupuncture anaesthesia? These questions are of great interest to the medical circle. At the National Symposium on Acupuncture and Acupuncture Anaesthesia held in Beijing in 1979, over 500 papers were read. They summarized and discussed the results of clinical application as well as theoretical study in this field. There are three major theories that attempt to explain the effectiveness of acupuncture. The meridian theory, or channel-collateral theory, is based on the conception that there is a free flow of 'qi' (vital energy) through the meridians of a healthy body. This flow is governed by the interplay of two complementary opposites, the 'Yin' (negative) and the 'Yang' (positive). Disease or pain results from their imbalance. Needling appropriate points in the meridians may restore balance and cure the ailment or relieve the pain. The Shanghai Institute of Physiology with the cooperation of other research units has made tremendous contributions towards the elaboration of this theory. A large amount of experimental work on animals has revealed that needling certain points may produce notable regulating action on the functions of the respiratory, circulatory, digestive, endocrine, nervous and energy metabolism systems. Acupuncture therapy also shows a remarkable enhancing influence on the immunological functions of human blood cells.

Another model, the neurophysiologic theory, is built around the idea of 'gate-control' of pain. Acupuncture is thought to prevent pain sensation from reaching the brain. The Shanghai Institute of Physiology has shown that neurons in a certain area of the thalamus discharge characteristic patterns of impulses in response to 'painful stimuli'. These responses can be suppressed by morphine or significantly diminished by electrical needling of particular acupuncture points. Thirdly, some researchers in Shanghai First Medical College and Shanghai Institute of Physiology held up the humoral theory. They suggest that certain chemical agents of an analgesic nature (morphine-like substances) were produced by the needling. They have noticed that the pain threshold of rabbits can be increased by the injection of subarachnoid fluid or blood from rabbits treated with acupuncture. The relationship between this endogenous morphine-like substance and the analgesia resulting from acupuncture has aroused considerable interest.

The integration of traditional and Western medicine is one of the guidelines in health care policy. The main goal is to build up a new kind of Chinese medicine, distinct from and superior to either traditional or Western medicine. In November 1981 the Chinese Association for Integration of Traditional and Western Medicine was set up. Some progress has already been achieved since the 1960s. For example, the prognosis of some refractory diseases has been greatly improved after treatment methods combining traditional and Western medicines became available. Also, Xi-yuan Hospital under the Chinese Academy of Traditional Medicine has done research on the treatment of chronic aplastic anaemia with combined traditional and Western techniques. Aplastic anaemia is a syndrome of pancytopenia caused by marrow failure. The usual treatment consists of taking testosterone, adrenocortical hormone or cobalt chloride or performing splenectomy, but the prognosis used to be poor. The specialists in Xi-yuan Hospital prescribe Chinese herbs to strengthen 'the kidney' and 'the spleen'. As a result the recovery of haemoglobin and leucocyte levels improved; the effective rate is 70 per cent.

The orthopaedists and osteologists of the Tianjin Municipal Hospital have treated fractures with integrated traditional and Western medicines. The main principles of the method are the combination of exercise and immobilization, treatment of the patient as a whole, and simultaneous attention to both muscular and skeletal components. The procedure is adaptable to over 90 per cent of various kinds of fractures involving different parts of the body. The non-union rate is only 0.4 per thousand.

4   Agriculture and Environmental Protection

Science and technology relating to the use of land for growing economic crops and to the management of the environment are the subjects of discourse in this chapter. Farming occupies a vital place in China; it is the principal occupation of over four-fifths of her working populace and remains the main determinant of the general health of her economy. In broad perspectives she is doing reasonably well in agriculture. With only 7.3 per cent of the total arable area but almost a quarter of the total population in the world, she manages to be nearly self-sufficient in food production. Comparatively, while in terms of per capita cropland she and the UK about stand equal, the latter depends as much on imports as on home supplies in feeding the nation. This observation should be balanced by the realization that only one-fortieth of the British labour force toil on the land. Moreover, agricultural expansion in China is limited by a number of objective factors as will be discussed in the following paragraph. It has continuously lagged behind industrial development, its share of the GNP now being a mere sixth: refer to Table 3 in Appendix 1. While the increase in the output of cash crops has been rapid especially during the last several years, that for staple crops has been modest and in fact is barely ahead of the rise in population. Owing to urbanization, soil destruction via salinization and alkalization, and desertification (see section 4), the total grain-arable land is shrinking. Despite the improvements in the multiple cropping index (pp 61–2), the aggregate grain-sown area is also decreasing, especially between 1978 and 1981, when the drop was by as much as 6m ha. At present only about 50 Mt of 15 per cent of the annual grain harvest are purchased by the state and made available for urban consumption. Furthermore, due to the inadequacy of transportation, shortfall in commodity grain can happen not only on the national but also on the local level. Indeed, in a sense the whole situation may be summarized by two long-standing directives, adopted in 1961 but still valid, viz. 'Agriculture is the foundation of the economy' and 'Take grain as the key link'.

Chinese scientists have in past decades done much in the field of agronomy. Examples of notable progress are genetic improvement of wheat and maize by selection, breeding of hybrid rice and allocotoploid triticale, discovery of the winter migration of armyworms, prevention and control of locusts and wheat midges, as well as biological control of many major plant diseases. Successes were also achieved concerning rabbit-attenuated vaccine against swine fever and rinder pest, diagnosis of infectious horse anaemia and acquisition of its immunity, etc. Nevertheless, there is an urgent need to further strengthen research efforts. Unless fundamental breakthroughs are sought and attained in agricultural science and technology China cannot easily boost her crop harvest drastically. The reasons are the scarcity of farmland almost all of which is under intense cultivation and much of it supporting multiple cropping, the diminishing return from labour inputs which to a considerable extent compensate for the moderate degree of mechanization, and the relatively high level of farm productivity already reached with the use of improved strains and of natural and chemical fertilizers. It is therefore not surprising that agriculture tops the list of eight priority disciplines delineated in the National Plan for Scientific and Technology Development, that was originally formulated for the period of 1978–85 and is detailed here in Appendix 6. Additionally, among the other seven is genetic engineering which includes the breeding of crop varieties. In 1982 China took out a US$60m loan from the World Bank to finance an agricultural development project (p 79), and a $75.4m sum from the same source in the following year to fund an expansion in agronomic research and training facilities, for which purpose the government itself will spend an additional $125m. The two loans were the second and the third from the international organization; the first, negotiated in 1981 and amounting to $200m, was channelled to the purchase of equipments related to higher education in general, towards which end the country contributed $95m herself.

The first two sections of this chapter will examine the sciences of crop breeding and protection, both vital aspects of the effort to guarantee good harvests. The third will deal with the technologies of chemical fertilizers production and irrigation. Fertilizers are essential to the intensive cultivation practised by China; the importance attached to irrigation may be gleaned from the Chinese slogan: 'Water conservancy is the lifeblood of agriculture'.

Animal husbandry supplies only 14 per cent of the national agricultural output in terms of monetary value and, as a research field, has attracted much less attention than crop farming. On the one hand, livestock production, which remains largely in the private as opposite to the state or collective sector of the economy, cannot be raised significantly because of a shortage of grain. On the other hand, cattle rearing is limited by the grazable acreage, totalling 270m hectares (224m for grassland plus 46m for hillside pastures), and by the underdevelopment in transport and refrigeration which severely hampers the shipment of meat to densely inhabited cropland. Furthermore, the percentage of animal proteins in the average diet already suffices to maintain a healthy nutritional balance in China; it is wise to learn from the negative example of the West, where the proportion consumed is excessive. In this chapter animal science is to be omitted from discussion. The subject of fishery will be touched upon, however, in the next chapter where the Institute of Hydrobiology is more fully described. Forestry is examined in section 4, being important from the points of view of not only economic resources but also environmental protection. The last-mentioned sphere of activities, in which China has accomplished much during the past several years, will be looked at in greater depths in the last section. The present emphases on the areas covered in these two sections are reflected by the fact that, in 1978, a Ministry of Forestry became detached from the Ministry of Agriculture and, moreover, survived the administrative reform of the whole government in 1982, when many ministries were abolished, merged or downgraded to bureau status, but then a new ministry was created with 'Environmental Protection' included in its title.

4.1 Crop Breeding

In order of decreasing shares of production, grains sown in China comprise rice, wheat, maize, potatoes and others. (Potatoes are grouped under 'grain' there; the official figure of weight equivalence for this tuber is 0.2.) The main oil-bearing crop is peanut; soya beans give a greater output but again are put in the category of grain because of their high protein contents. Of the inedible crops cotton is the major representative. Of the over fifty species cultivated as vegetables, the four leading varieties, in terms of acreage, are tomato, cucumber, Chinese cabbage and cabbage. Generally speaking, compared to the other groups, vegetables and fruits receive the least attention regarding genetic breeding.

In the last three decades Chinese scientists have evolved some 3 000 cultivars of 41 economic plants including grain, oil-bearing, sugar and fibre crops and tobacco. As said before grain occupies the central place in food supply and, naturally, research has been focused on rice and wheat, the former grown more in the south and the latter mostly in the north. Since 1949 the two have their dominant cultivated strains changed three times nationwide and many more varieties have been popularized on a large scale.

For the purpose of raising production the rice area is being expanded to the limits of adaptation, towards the colder and drier northeast and the mountainous west, and now extends to some 35m hectares. Short-statured, high-yielding cultivars have been bred well before those by the International Rice Research Institute. Moreover, the Chinese strains require shorter growing periods than those first released by the IRRI, namely IR-5 through to IR-26. In China, early maturity comes close to being the universal objective of any plant selection programme due to the desire for greater productivity through multiple cropping. A complex cropping pattern has already emerged that includes double, triple and relay cropping as well as intercropping. These practices, which are necessitated by the scarcity of arable land, demand exact timing, whereas earliness will minimize the adverse influence of changeable weather. The national mean cropping index has now reached about 2. Even at the latitude of Beijing, ie 40 degrees N where the growing season lasts only 160 days, rice preceded by wheat is a common sequence. Examples of intercropping found in various provinces are wheat-soya beans, wheat-corn, wheat-vegetables and wheat-cotton. In 1908 China's per-hectare yield was, for rice, 4 120 kg on average, wheat 1940, soyabeans 1 030 and corn 3 095 kg. The main reason for these figures being modest compared to the best in the world is the high proportion of land under multiple cropping and intercropping cultivation.

Winter wheat, grown in the more temperate parts of the north, accounts for some 85 per cent of total wheat output although spring wheat, traditionally cultivated in the northwest, is gaining in proportion as relay and inter-cropping patterns that include it spread southward. This trend has, as said before, been possible only because of the availability of varieties with shorter life cycles. Semidwarf high-yielding types were evolved in the past decade, through the use of indigenous dwarfing genes as well as dwarfism from Korean, Mexican, East European and other sources. With the acquisition of insensitivity to day length as the aim, a number of winter × spring crosses have also been made: the successful outcome has allowed the early planting of spring wheats in the southern provinces.

After wheat, maize ranks third among cereals grown in China. The hypothesis has been formulated that it is a native crop, originating from Yunnan province, but the general view holds that it was from North America and first introduced in the beginning of the 16th century, and that later it mutated (Wx into wx) independently in both China and America. The number of original introductions was probably small and, as in the US today, the genetic heterogenity of cultivated maize is low, although the species reproduces by cross-fertilization. A programme of producing inbreds from double-cross hybrid seeds was initiated but suffered a severe setback in 1966, when an epidemic of northern corn leaf blight (Helminthosporium turcicum) swept through the northern regions. In that year, under the impact of the Cultural Revolution, the organization of plant breeding was just beginning to change radically to involve the masses. Resistant strains were subsequently identified, mostly from local open-pollinated varieties, often by peasants. Since then the work shifted towards single-cross hybridization to preserve the resistance. Chinese delegates to an international conference in 1977 stated that in the main maize belt of North China, 55 per cent of the area in hybrids was in single crosses, 40 per cent in double cross and 5 per cent in three-way or top-cross materials.

Mutation breeding is commonly undertaken in the country, especially since mutational defects usually result in early maturity and, with the broad masses mobilized to take part in the job of selection, a large number of mutants can be tried. Mutagenic agents utilized include fast neutrons, X-ray and UV light; rather surprisingly, chemicals such as sodium azide are seldom employed but gamma radiation, frequently. In the last few years mutations caused by energetic electron irradiation have also been reported in the literature. Species treated are rice, wheat, maize, soya beans, millet, Chinese sorghum (in Chinese 'gaoliang' or literally 'tall millet'), rape, cabbage, tobacco, yeast and bacteria, among others. For its work on mutation-bred cotton the provincial Shandong Cotton Research Institute won a state award in 1918. Named 'Lumian 1', the new strain was created in 1974 through gamma irradiation by cobalt-60, and has a productivity 25 per cent greater than the 'Deltapine 15' from America, achieving a mean per-hectare yield of ginned cotton of 1 970 kg.

Hybridization techniques in China are characterized by an uniformly high standard, having been applied to all the other major crops besides wheat and maize. The programme on rice, started in 1964, was a difficult one since this species propagates by self-fertilization. By 1973, however, the scientists responsible succeeded in obtaining the 'three lines' – a Chinese term referring to the complete system of male sterile, maintenance and restoration lines. For this achievement the researchers were given an award by the State Science Commission in 1981. Between 1976 and 1980 17m hectares of land were planted with the new hybrid. Easy to till and suitable for a wide range of climates, the cultivar has large root system (hence the resistance to drought), strong stems and large ears with quality glutinous grains (having 9–11 per cent protein content). It has been the subject of technological transfers to ten foreign countries, the latest deal being with Ring Around Corporation, a subsidiary of Occidental Petroleum. Chinese agronomists proved its adaptability to the conditions in the southern United States, where field trials proved that its unit yields consistently exceed the respective yields of the US's own Starbonnet strain.

The procedures for emasculation and pollination in hybridization bear similarities to those practised abroad. The florets are cut back, anthers together with pollens carefully removed, and plastic bags placed enclosing the emasculated spikes. Fertilization is realized either through the approach methods or by cutting the top of the plastic bag in question and twirling the pollen spike until its pollen is shed onto the emasculated flowers. The technique for handling segregating populations follows the pedigree system for the most, although some modifications may be imposed in later generations depending on the objectives of specific research projects. Yield evaluation of the fifth or sixth and later generations is normally conducted on the commune level. The familiarity with agronomic trait measurements required of the common peasants comes from books, posters and propaganda by 'barefoot' agronomists.

Haploid breeding is an activity in which a remarkably broad spectrum of the people are involved. China is alone in having popularized this methodology with tremendous enthusiasm. In addition to professionals at universities and institutes' peasants at hosts of the field stations run by communes try their hands on anther and pollen culture. The work commenced in 1970 and success with wheat came first, in 1972: the Institute of Genetics could justifiably claim to be the first in the world in achieving wheat reproduction through pollen culture. Hot on the heels came the success with tobacco. It was shown that calluses were readily cultured from the pollen and without great difficulties induced to grow into plantlets. The case of rice also proved relatively easy. Subsequent refinement of techniques such as cold pretreatment and 'fine tuning' of culture media enables many new varieties to be derived from doubled haploids for an array of crops encompassing maize, rape, sugar cane, red pepper, citrus (Microcarpa bunge), aubergine, flax (Linum usitatissimum), rubber tree, poplar as well as triticale (wheat × rye) and wheat × Agropyron hybrids. The methodology has been applied in as diverse a context as forestry (section 4) and medicine. The Institute of Traditional Chinese Pharmacology has demonstrated that 'Qinghaosu' (see section 7 of the previous chapter) may be mass-produced via callus induction and plantlet regeneration of Artemisia Annua L.

The culture medium now usually used for rice calluses is Miller's solution with about 80 g/1 sucrose and 2 mg/1 of 2,4-D, whereas for wheat it is often Murashige and Skoog medium plus various supplementary components. Anthers with pollen at the early to mid-mononucleate stage are taken from the F1 plant and seeded on agar containing the medium. Callus formation frequencies in wheat are improved when the spike has been chilled before pollen removal, but remain lower than in rice. Plantlets thrive in a medium containing 0.2 to 10 mg/1 of 1AA and 1 to 2 mg/1 of a kinetin but a reduced percentage of sugar. They may next be transferred to a third medium depending on the species. Finally they are reverted to diploids through either spontaneous or colchicine-induced chromosome doubling.

Experiments with ovary and ovule culture started in 1978. Haploid tillers have been obtained in the cases of Nicotiana tabacum and Triticum aestivum at the Institute of Genetics.

Somatic hybridization is, like ovary culture, in an early stage of development. Whereas somatic cells of tobacco and later rice have been successfully regenerated to plantlets, callus tissue has not been produced from any other species. Much of the current effort focuses on the somatic embryogenesis of ginseng (Panax ginseng Meyer), a precious plant in Chinese traditional medicine. There are not only internal but also international competitions in this work, from Taiwan, as well as from the Soviet Union. As in other countries the molecular biological approach to realizing genetic variability is regarded as an exciting possibility. Genetic engineering was listed among the eight research fields assigned top priority in the original 1978–85 National Plan for Scientific and Technological Development, which specified that it should be applied to 'evolve new high-yielding crop varieties capable of fixing nitrogen'. However, no major work in this particular direction has yet been reported.

We now take a look at the organization of crop breeding research. Formal research units are set up on the national as well as provincial, county or even commune levels. Indeed, as said before a unique feature of Chinese breeding programmes is the meaningful participation by the masses. Nevertheless, attention will be restricted to those administered by a ministry either on its own or jointly with a provincial or municipal authority. Units under exclusive local control are not only too numerous to cover in a few pages, but also liable to be given descriptions that contain inaccuracies or gross omissions, as up-to-date information on them is generally less accessible.

Seven institutes under the Academy of Sciences, which enjoys ministrial status, can be identified as supporting crop breeding projects currently. The Institute of Genetics, in Beijing, has already been mentioned in previous paragraphs. Its existence dates from 1959 but its predecessor, the Experimental Establishment of Genetics and Plant Breeding, was founded in 1951. It now has six divisions, devoted respectively to investigating molecular genetics, animal and human medical genetics, plant cell genetics and cell engineering, applied genetics, cytoplasmic genetics and the interaction between cytoplasm and nucleus, evolution genetics, as well as novel methods in genetics and breeding. It also runs an experimental outstation for cotton research on Hainan Island. The work on genetic engineering there is most advanced in connection with microbiology. Some recent accomplishments in crop breeding are: ovary and ovule culture of tobacco and spring wheat; anther culture of rice, wheat, maize, sugar cane and rubber tree; isolated pollen grain culture of maize and sorghum; and overcoming incompatibility and sterility in distant hybridization.

The Institute of Botany, also in Beijing, is composed of nine divisions namely: Plant Taxonomy and Geography, Plant Ecology and Geobotany, Phytomorphology, Plant Physiology, Photosynthesis, Photochemistry, Nitrogen Fixation, Plant Cytology, and lastly Palaeobotany. Besides the study of rejuvenation of nuclei and haploids by anther culture, some other contributions from the Institute are the cloning of cold-resistant grapes and the cultivation of new crops in the Beijing region through plant introduction and acclimatization. One of the tasks for the South China Institute of Botany, at Guangzhou, is similarily to introduce and acclimatize economic crops from tropical and subtropical zones. Among its seven divisions is one for Plant Genetics, where crop breeding is underway with the focus on medicinal and oil-rich plants. An outstanding success is the clonal propagation of Begonia fimbristipula Hance, a species highly valued in Chinese traditional pharmacology. Calluses from epidermal cells of aseptic seedling leaves were induced to differentiate into plantlets on Murashige and Skoog medium supplemented with 0.25 ppm of KT, 0.1 ppm 2,4-D and 2.5 ppm NAA. An existing project is to grow haploid plantlets of Brassica campestris. To the west of Guangzhou still within the subtropical zone is Kuming, the location of another botanic institute. The three divisions of Plant Taxonomy and Floristic Geography, Plant Physiology, and Phytochemistry constitute this Kuming Institute, to which a group is also appended that specializes in haploid breeding.

The Chengdu Institute of Biology, formerly the Southwest Biological Institute before renamed first the Agrobiology Institute, comprises seven divisions. Their disciplines of specialization are respectively bio-energy resources, amphibians, botany, microbiology, cells, molecular genetics, and crops genetic breeding. In the first division there is work on biogas production, of which more in Chapter 7.5. Present efforts at the last division are directed towards wheat, rape, and double-harvest rice in the Chengdu plain. The focus of research in the Northwest Plateau Institute of Biology, Xining, is on biological problems peculiar to farmland of high altitude. Several strains of spring wheat have been created via hybridization and anther culture techniques, and work is in progress to determine the hereditary variability of their characters. Lastly, the Shanghai Institute of Plant Physiology is interested in cell fusion. Interestingly, despite its title, it is initiating a programme on the molecular genetic control of microbial production of antibiotics. It runs a phytotron which extends to 360 square metres in area, has a temperature adjustable from 0 to 50 degrees C and a humidity 30–90 per cent, and is used by visiting scientists from all parts of the country.

Before leaving the Academy of Sciences, our attention should touch upon the Institute of Nuclear Research, in collaboration with which many plant breeding organizations have carried out their mutation experiments. Located in a county within the Shanghai municipality, the Institute houses a cobalt source stored in a water-filled well. Also available there is a neutron generator driven by an eight-step cascade transformer affording 150 to 200 kV output. Deuterium is fed through a high-pressure palladium leak into a RF ion source, which can provide an ionic current of 3 to 5 mA. With tritiated targets the total flux integrated over the entire solid angle reaches 10^{10} neutrons per second. An overall description of the Institute can be found on p 162.

The Academy of Agricultural Sciences is the research arm of the Ministry of Agriculture, Animal Husbandry and Fishery. Individual institutes known to be in the Academy carry the respective designations:

Atomic Energy Utilization in Agriculture, located in Beijing
Crop Breeding and Cultivation, Beijing
Germ Plasm, Beijing
Rice, Tianjin
Oil Crops, Wuhan
Tropical Crops, Danxian (Hainan Island, Guangdong)
Olericulture, Tianjin
Citrus Fruits, Chongqing
Pomology, Xingcheng county (Liaoning)
Pomology, Zhengzhou
Cotton, Anyang (Henan)
Jute and Kanaff, Yuanjiang county (Hunan)
Tea, Hangzhou
Sugar Beet, Hulan county (Heilongjiang)
Tobacco, Yidu county (Shangdong)
Rubber Tree, Wenchang (Hainan Island)
Plant Protection, Beijing
Biological Control, Beijing
Natural Agricultural Resource Survey and Agricultural Regions
    Research, Beijing
Agricultural Economics, Beijing
Soil and Fertilizer, Beijing★
Agricultural Mechanization, Harbin
Farm Field Irrigation, Xinxiang (Henan)
as well as
Sericulture, Zhenjiang
Apiculture, Nanchang
Grassland, Xilinhaote (Inner Mongolia region)
Animal Schistosomiasis, Shanghai
Animal Husbandry, Lanzhou
Veterinary Science, Lanzhou
Veterinary Science, Harbin

★In case confusion should arise: until a couple of years ago the Institute of Soil and Fertilizer was situated at Dezhou in Shandong province, but since then it has moved back to Beijing, where it had been before the Cultural Revolution.

Chinese Traditional Veterinary Medicine, Langzhou
Information of Agricultural Science and Technology, Beijing

The Institute of Atomic Energy Utilization in Agriculture is made up of three laboratories with the respective titles of Instrumentation, Isotope Tracing, and Plant Breeding. The last one possesses a cobalt-60 source immersed in a pool of water inside a shielded vault. Seeds to be irradiated are left on top of the pedestal inside the vault and the source is lifted up. Both radiation-induced mutation and its combination with hybridization have been attempted in wheat, wheat $\times$ rye, wheat $\times$ sorghum and other plants. A genetic-cytoplasmic sterility system has been established for sorghum. The Institute of Crop Breeding and Cultivation has crossbred Chinese and imported strains of rice, wheat, cotton, kenaf and others. Some genetic breeding programmes are also underway in those institutes which are oriented towards specific groups of crops. Efforts at the Citrus Fruits Research Institute has borne fruits in eg hybrid embryo culturing of sweet oranges. The Tobacco Institute has evolved a new tobacco cultivar, a monoploid which gives higher yield and greater resistance to diseases. The South China Tropical Crops Research Institute at Danxian has succeeded in winter pollination of sugar cane and obtained several improved varieties via hybridization.

In additional to the national organization depicted above, there are local academies of agricultural sciences in virtually all of the country's provinces and autonomous regions and in the three municipalities. In the beginning these academies functioned as branches of the Academy of Agricultural Sciences, but since the Cultural Revolution they have all been transferred to the jurisdiction of corresponding local governments. As an example of the work at their research centres, molecular hybridization is being applied at the Institute for Economic Crops, Nanjing, that belongs to the Jiangsu Academy. In one series of experiments, DNA extracted from the seeds of sea-island cotton '416' (Gossypium barbadense) was injected into fertilized bolls of upland 'Glandless' (G. hirsutum), and character variations were analysed down to the third generation in the offsprings. Other institutes operated by the Jiangsu Academy carry the respective designations: Grain Crops, Plant Protection (see p 69), Soil and Fertilizers, Agricultural Physical Chemistry, Horticulture, Animal Husbandry and Veterinary Science, and Aquatic Products. The Institute of Rice under the neighbouring Zhejiang Academy is looking at somatic breeding. Many sets of uniform progenies have been obtained, and systematic studies on D1-D5 generations have been completed for several cases of paddies. This institute is situated at the provincial capital of Hangzhou, which hosts also the fellow Institute of Soil and Fertilizers, of which more on p 78. Shanghai municipality at the eastern corner of the boundary between Jiangsu and Zheijiang set up, in 1960, its Academy of Agricultural Sciences. The Academy now contains five institutes: Plant Breeding, Plant Protection (p 69), Soil and Fertilizers, Agricultural Sciences, Horticulture, and Livestock and Veterinary. The first institute is experimenting with rice anther culture; a number of crosses have been observed down to the F5 generation.

Plant breeding is carried out in some institutions of higher education, which in this context fall into three categories. The first embraces those that are controlled by the Ministry of Education. An illustrative piece of recent work in such places is the study, at the Biology Department in Beijing University, of DNA synthesis in Brassica pekinensis pollen during the initial period of anther culture. The method employed was micro-autoradiography involving tritium-TdR labelling. The department in Wuhan University is studying the test-tube fertilization of paddy rice, Oryze sativa L. That in Southwestern Normal College at Chongqing has investigated gamma-ray induced chimaera of awned mutation in the M1 generation of Triticum aestivum L. The department in South China Normal University, Guangzhou, recently pioneered the use of the electron beam as a mutagenic agent in rice seeds. The source was the BF-5, a China-made 3.5-MeV linac capable of providing a sustained current of 0.2mA. Electron irradiation was found to result in lower damage, higher mutation frequency and wider mutation spectrum than gamma-ray irradiation.

The second category refers to those under the dual leadership of the Ministries of Education and of Agriculture, Animal Husbandry and Fishery. An example is the Northwest College of Agriculture at Wugong, which is one of the key centres for research on winter wheat. The first semi-dwarf variety was released from this college: labelled I-fung No. 3, it incorporated the dwarfing gene from the Suewon No. 86, a Korean strain imported in mid-1970's. In the Beijing Agricultural University, a group from the Agrobiophysics Department is examining the genotypes of the progenics of Hordeum vulgare seeds treated with various mutagenic agents. The Sichuan Agricultural College at Yaan has conducted hybridization breeding of barley such as Hordeum distichon L. × hexatichon L. The Jiangsu Agricultural College, Yangzhou, is looking at the genetic variance and heritability of characters in the F2 population of wheat crosses. The third category comprises institutions run by local authorities. These schools by their nature were not bestowed the status of 'key' universities and, generally speaking, their emphasis is more on teaching than on research.

4.2 Disease and Pest Control

Agricultural protection is defined in China as 'integrated control with prevention put first'. It is a combination of extension and set-purpose research in the two subjects of phytopathology and entomology, supported by lesser efforts in basic research. As in the case of crop breeding such studies are carried out on many levels. National and provincial institutes number in the tens. Additionally, most communes maintain laboratories furnished with simple analytical instruments as well as reference collections of common diseased plant and insect pest specimens. Facilities for culturing organisms used in the biological control of insects sometimes also exist there. Visiting American scientists have gauged the total count of plant protectionists in the country to exceed that anywhere else, and judged Chinese protection measures to be outstandingly effective on the whole.

Genetic breeding with the goal of enhancing disease resistance is undertaken in most of the institutes discussed in the previous section. The case of northern corn leaf blight in maize has also been mentioned there. For vegetable crops, the selection by peasants over the centuries has likewise led to lines with high levels of resistance. Indeed, in the case of cucumbers, the primary source of mosaic resistance relied on extensively by American plant breeders is the 'China Long' cultivar introduced from China in 1924–5. For wheat, lines with high resistance to common bunt, and stem, stripe and leaf rusts have been bred. Although nearly every main disease that infects wheat is known in the country, no major epidemic has broken out since the one involving stripe rust in Shaanxi in 1964. The Chinese now possess well over 600 varieties resistant to stripe rust, which is moreover suppressed by the practice of timely planting. The wheat maggot used to cause extensive damage to spring wheat, before the availability of resistant cultivars such as 6407 and 6410. These varieties also tolerate stripe rust and soil alkalinity.

The only disease of economic significance not yet under control is scab. Its threat is especially serious in the winter wheat region from Xian eastward to Jiangsu province: wheat-corn rotation is a common practice near Xian and favours the spread of the scab fungus, whereas rice in Jiangsu similarly acts as a host. Sources of resistance to it have been identified in scores of cultivars including Ming Wheat No. 3 and Jin Guang, apart from Nung-lin No. 50 from Japan, Vanissa from Austria and Frontana from Brazil. Confirmations have been made that the resistance is genetic rather than simply due to early flowering. An acceptable, high-yield strain for the scab areas should be available soon. In general, the Chinese usually evaluate resistant varieties by exposing them under natural field conditions to the pathogen in question. At the Plant Protection Institute operated by the Jiangsu Academy of Agricultural Sciences, at Nanjing, resistance to wheat scab is being measured. The scab fungus is cultured on autoclaved barley seeds which, after perithecia are formed, provide the ascospore inoculum required for uniform infection of susceptible lines.

In the absence of scab resistance in current lines, fungicides are relied upon. The one of common choice is 1-methyl-1H-benzimidazole-2-carbox-amide. The wheat is sprayed once at about flowering, and often a second time a week later. The Plant Protection Laboratory under the Shanghai Academy of Agricultural Sciences has proposed a more effective procedure entailing pathogen forecasting, after studying release patterns of sexual spores. Spore traps are installed and readings taken regularly by 'bare-foot' agronomists. When the count of trapped ascospores exceeds seven in a microscope field of view at 400× the systemic fungicide is applied. This procedure has proved effective in trials by communes near Shanghai. The Institute is also experimenting with different seeding dates which affect time of flowering and thereby disease potential. One more way of controlling scab consists in good drainage. Perithecial production is reduced by keeping the top soil as dry as practical.

Indeed, in general the Chinese favour 'integrated control' in which the cultural approach is basic and chemical means are only supplementary. For vegetable crops early or late planting to avoid particular disease is also a common measure. It is used, for instance, against Phomopsis, Phthium and

Phytophthora root rots in eggplants and peppers, mosaic diseases in Chinese cabbage, i.e. Brassica pekinensis, B. chinensis and B. parachinensis, and wheat yellows caused by the barley yellow dwarf virus (BYDV) and the wheat rosette dwarf. Extensive ditch and furrow irrigation in China has resulted in the relatively infrequent occurrence of certain bacterial disease of the Chinese cabbage, legumes, cruciferae and solanaceous crops. At the same time, the practice of flooding paddy rice fields for two to four months each year helps, by maintaining anaerobic conditions in the soil, to keep in check the abundance of aerobic soil-born pathogens.

Fungicides remain indispensible in the combat against many foliar pathogens. The Chinese discovered gingfengmycin, an antibiotic which can be extracted from a species of Streptomyces and which acts against rice blast. They have simplified its extraction process so that it may be produced on the commune level. They also initiated the use of 'antimycin 120' against wheat rust and powdery mildew on cucumbers. Bordeaux mixture, often prepared immediately in the field, is still used to treat foliage blights. However, in recent years the dithiocarbamates are playing a greater role in the control of downy mildews on vegetables and curcurbita, leaf mould, and early tomato blight; organic fungicides in general have become more important. A second widespread application of fungicides is for seed treatment. In fact, treating seed with chemicals for disease prevention is a traditional method dating back to well before 1st century BC. In the country today, all seed lots sent across a county boundary have to be disinfected, and quarantine restrictions are enforced without exceptions on all agricultural produce shipped from abroad. Apart from flotation in chemical solutions, selection of parent plants and cell culture techniques such as stem tip culture in the case of potatoes are adopted to obtain seed from pathogens.

Unlike fungicides, herbicides are used much less than in the West. Chinese peasants are not adverse to the commitment to remove weeds by hand in exchange for the advantage of reducing chemical pollution in the field. However, chemicals in the form of insecticides still occupy a crucial position in China's plant protection programmes, by serving as supplements to cultural and biological controls. One of their roles is to destroy carriers of epidemics. As an example, Barley Yellow Dwarf Virus (BYDV) is responsible for probably the most serious viral disease of wheat in the country and it is transmitted by five or more species of aphids. The growing prevalence of inter-cropping favours it by allowing higher mobilities of the aphid. The main preventive method is therefore the spraying of insecticides such as Thimet, an organic phosphate, to kill the vectors. The other role is to function directly as pesticides against herbivorous insects and arachnids. Over ninety species live off rice, the most harmful among which are the rice stem borer, green rice leafhopper and brown planthopper. Changes in cultural practice towards double and triple cropping have decreased the severity of attack by the first insect but increased that by the second; the third is moreover a vector of viral diseases (yellow stunt, yellow dwarf and common dwarf). For the extermination of these insects rice beds are sprayed with DDT, fenitrothion or dimethoate emulsion. Chemical insecticides administered to early rice fields are mostly mixtures of BHC and methyl parathion, whereas fenitrothion and trichlorfon are used additionally in later times when the rice matures.

Cultural control methods are widely implemented. Flooding of rice fields in early March drowns larvae of rice paddy borers. The late rice may be protected by weeding, with leafhoppers and planthoppers destroyed in overwintering sites. Flooding to a depth of 8 cm when the rice plants reach a height of 20 cm forces mobile pests like planthoppers to the tips of the plants. Ducklings are then released at a density of 300 per ha to eat the insects. Other biological ways of control are also commonly practised. The population of rice stem borers is reduced by chalcid wasps. Twelve species of Trichogramma have been reported, of which four, namely japonicum, confusum, ostrineae and dentrolimi are mass-reared to be released in the field to prey on rice leafrollers, stripped paddy, corn, sugar cane and internodal borers, cotton bollworms, spiny pine caterpillars and other lepidopterous pests. Rearing involves the use of eggs of the rice grain moth or the giant silkworm as hosts; one egg of Antheraea pernyi produces on average some 60 Trichogramma. Depending on the target pest the degree of parasitization is usually 60 to 80 per cent. Efforts to popularize the utilization of this hymenopterous parasitoid as a 'living insecticide' include the distribution of posters and film shows in rural areas. A third kind of wasp, Dibrachys cavas, parasitizes overwintering pink bollworms (Pectinophora gossypiella) and is applied to cotton seed.

Several strains of the entomogenous bacterium Bacillus thuringiensis are used on a large scale against lepidopterous larvae in rice, cotton, maize, vegetables and forests. The strain B. thuringiensis ostrineae, discovered and characterized by Chinese microbiologists, is the best regarding exotoxin production. The application of the fungus Beauveria bassiana to maize during spring and autumn is made in a more limited way to kill the European corn borer and its overwintering larvae. Nuclear polyhedrosis viruses (NPV), granulosis viruses isolated from the Chinese cochlid and the cabbage moth, and a cytoplasmic polyhedrosis virus from the masson-pine caterpillar have been tested for insect control. NPV is found to be lethal to tussock moths on mulberry. While on the subject of sericulture we may further note that China is one of the countries that pioneered the use of juvenile hormones (insect growth regulators), initially by Shunde county of Guangdong during the early 1960's.

In the same province, entomologists from Zhongshan University at Guangzhou recently discovered a NPV strain to be deadly to the cotton leafworm that lives off various vegetables and legumes. They have thus opened a way for reducing the dependence on hazardous insecticides by which alone the pest could be controlled in the past. Earlier, their colleagues had found the eupelmid egg parasite, Anastatus sp., to be effective against the stinkbug on litchi fruit trees. Also, another group furnished a scientific study of the traditional Chinese practice of herding ducklings through paddies. The extent was measured to which rice pests, both insects and weeds, as well as vectors or alternative hosts of human diseases could be eradicated.

Plant breeding for resistance to insect pests receives less emphasis than that for disease resistance improvement. One circumstantial factor is the absence of interdisciplinary research units combining genetics and entomology. Nevertheless, advances have been made in the last two decades. Rice species relatively insusceptible to sheath blight (Pellicularia sasakii) have been bred by the Guangdong Academy of Agricultural Sciences. In the winter wheat region to the east of Xian, the once serious threat posed by red sucking flies (Sitodiposis mosellana) was eliminated by the planting of a resistant line. The line was evolved at the Shaanxi Academy through selection from survivors of a native variety in a formerly heavily infested area. A soybean type, the Keling No. 3, is extensively grown after having been identified by the Jilin Academy as being avoided by the soybean pod borer; resistance to aphids, the other group of major soya bean pest, is still being sought after. In the case of cotton, however, several strains have been selected that are resistant to Aphis gossypii Glover. Their availability is vital due to the mounting insusceptibility of the cotton aphid itself to organophosphorous insecticides. The mechanism of host plant resistance has been attributed to allomone secretion. A possibility being explored is the acquisition of aphid repellent characteristics by cultivated potatoes after being crossed with certain wild potato species.

Physical methods of insect control see greater utilization in China than anywhere else. Bait pails are placed at densities of 50 to 100 per hectare to catch insects in flight. The liquid contained is usually a mixture of 9 parts of water with 6 parts of sugar, 3 vinegar and 1 wine. Blacklight traps are also employed with spacings of one in 1 to 4 ha. A typical trap consists of a baffle enclosing a gaseous discharge lamp either in additional to, or in place of, an incandescent light the wattage of which is chosen for optimum emission spectrum corresponding to the target insects. Underneath is a basin of water covered with a film of DDT, kerosene or sticky rice water to immobolize the stunned insects. Of course, unavoidably both pest insects and, to various degrees, their predators and parasites are caught but, so long as the traps are in continual operation, the competing populations may stabilize at a low level. Blacklight traps can also serve the purpose of insect monitoring and forecasting, after modification whereby the water basin is substituted by electric grids or a jar of potassium cyanide. Population records have been kept for the past 30 years.

Insect forecasting is an essential component of integrated pest management, There is a hierarchy of forecasting organizations in the country. At the top of the system is the Central Forecasting Centre in Beijing, which maintains close contact with regional centres and the meteorological service. Provincial offices are responsible for long and mid-term work; short-term monitoring and local control are supervised at district and county levels. Teams from communes and production brigades implement the recommendations from above and at the same time feed back raw data. China for a long time has recognized the drawbacks of such uni-factor intervention as insecticide application, which include insecticide resistance after prolonged application, insufficiently narrow spectrum leading to indiscriminate depletion of beneficial insect population, the possible emergence of devastating secondary pests, and the danger that the toxicity is persistent resulting in crop residues and ecological contamination. Guided by the dialectical principle that all processes in the physical world occur not in isolation but can only be fully described by their inter-relations, the Chinese have rich experience in multi-dimensional control programmes integrating chemical, cultural, biological, physical and environmental approaches. Along this path, in terms of both general enthusiasm generated and concrete results achieved, China has outpaced everyone – a fact acknowledged by US entomology representatives visiting the country in 1975. The suppression of locust plagues, which knits together biology, ecology, hydrology, cultural practice and meteorology while relying heavily on bionomics, serves as a good illustration.

The earliest surviving record of locust control in the world is probably the use of fires mentioned in a Chinese poem of around the 13th century BC. The first anti-locust decree was issued in 29AD. A 17th-century book, proposed the planting of legumes, inedible to locusts, and the herding of ducklings through paddies. In the 1950s and 60s the aerial spraying of '666' (benzene hexachloride) was adopted but virtually abandoned by 1975. This curtailment was necessary, owing to pollution problems, and also timely, because environmental transformation measures was gradually taking effect. Thanks to the labours of a highly-motivated peasantry, numerous irrigation ditches, check dams and small reservoirs (the latter also functioning as fish ponds) had been built in infested areas to prevent flood surges. Potential locust-breeding grounds in reedy, riverside flood plains were converted into rice paddies or planted with shade, fruit or medicinal trees; the plantings create microclimatic conditions unsuitable for oviposition and egg hatching, and also expanded niches for frogs and other natural enemies of this pest. A rational system of measures for migration control has been formulated, based on knowledge of the phase polymorphism and gregarious behaviour of locusts. At present, a monitoring and forecasting network exists to maintain surveillance over possible local outbreaks.

Research relevant to plant protection is visible at several institutes belonging to the Academy of Sciences. Reference to the Institute of Genetics has already appeared in the last section. The Institute of Microbiology, also in Beijing, was founded in 1958 and has been active in the study of plant diseases. It is composed of eight divisions: Virology, Bacteriology, Mycology, Microbial Physiology and Ecology, Microbial Metabolism, Enzymology, Microbial Genetics, and Culture Collection. The Wuhan Institute of Microbiology (1956–78) or Virology (renamed thus in 1978) used to contain four virology departments. Work at the first of these continues to focus on virus classification and viral control of pests. Two more departments have been added recently: Agricultural Microbiology, where Bacillus thuringiensis and entomopathogens are studied, and Environmental Microbiology, where the biodegradation of petrolic and pesticidal pollutants is currently the main subject of interest. The pollution of crop and soil by heavy metals and chemical insecticides is also investigated at the Nanjing Institute of Soil Science.

The Institute of Entomology, established in 1959, now comprises the five divisions of Ecology and Taxonomy, Insect Physiology, Insect Toxicology,

Insect Virology, and Experimental Techniques. The method of controlling mulberry tussock moths with NPV, mentioned earlier, has been developed at this institute. Some other notable contributions are the identification of several toxicants for exterminating the termites Reticulitermes fleviceps and lucifugus santonensis, as well as the synthesis of gossyplure and 11-tetradecenyl acetate, the sex pheromones of the Pink Cotton Bollworm (Pectinophora gossypiella) and the European corn borer, respectively. An ongoing project concerns the paddy borer pheromone. Apart from combined electroautennography and liquid chromatography for pheromone studies, many ingeneous specimen preparation techniques for scanning electron microscopy were pioneered there, too. The Kunming Institute of Zoology was set up in 1958 and work was concentrated initially on vertebrate taxonomy and faunistics. It has, in recent years, branched out to other disciplines including entomology. Research in this domain is oriented towards the biological control of agricultural pests; one of the projects has been to clarify the regularities in the migration of aphids in Yunnan province.

Entomology is, also, a chief subject of investigation at the Institute of Zoology in Beijing. Five of the ten divisions there specialize respectively in: insect taxonomy and faunistics, insect ecology, insect physiology, insect pheromones, and lastly insecticides and toxicology. Significant progress has been made in the biological control of the spiny bollworm, cotton aphid, cotton leafworm (with the help of NPV), armyworm, pine caterpillar, etc. Some other achievements are the structural determinations and syntheses of Dendrolimus pheromones, in addition to the control of the oriental fruit moth. Concerning insecticides, it is the Institute which synthesized phoxim and sevin (carbaryl and completed their field trials. Here also it was discovered that the photodegradation of the supposedly safe phoxim gives rise to substances of high mammalian toxicities. An outstanding opus earlier on had been a detailed explanation of the migratory behaviour of locusts.

An example of work in the Academy of Agricultural Sciences, done at the Chongqing Institute of Citrus Fruits, is the biological control of Aleurocanthua stinosus kowayama. A wasp was propagated to act as the natural enemy of this pest of citrus trees. Another example is the solid fermentation process of culturing Bacillus thuringiensis, pioneered by the Soil and Fertilizers Research Institute now in Beijing. The Tea Institute has studied the degradation of phoxim on tea bushes under natural sunlight as well as artificially shaded field conditions; an unfavourable judgment was passed on this pesticide in view of residue toxicity.

The Plant Protection Laboratory of the Jiangsu Academy at Nanjing is investigating the viral control of bacterial rice blight, pathogen forecasting in paddy water and pathogen detection in crop residues. Three other illustrative pieces of research in the case of local academies have already been mentioned when host resistance breeding was discussed. The Shaanxi Academy at Wugong, besides developing a wheat variety resistant to midges, has also formulated a simple approach to the biological control of cotton aphids. Certain kinds of weeds are permitted to grow in the cotton in early season but pulled when the aphids appear, so that lady beetles thriving in the weeds are forced to the cotton shrubs.

Also at Wugong, the Northwest College of Agriculture, mentioned earlier at the end of the last section, has a plant protection department where the wheat cultivar resistant to red sucking flies, referred to in the above, was originally bred. Trials for field inoculation with Fusarium oxysporum f. vasinfectum on cotton have been performed at the college. Some of the staff are engaged in a long-term programme of evaluating pest management methods in the orchard. Deciduous fruit trees in Shannxi and surrounding provinces are threatened by some 120 species of pests or potential pests. The application in the 1950s of DDT at budding time to kill leafrollers in apple trees had let the spider mite to become the most serious pest. In the 1960s the trees were treated with organophosphorous compounds two or three times a year for diseases and seven or eight times for spider mites, but then the problem of insecticide resistance arose. In the case of peaches, spraying parathion during summer to control oriental fruit moths caused the emergence of San Jose scale. Since the mid 1970s the isolated use of pesticides was avoided. Plant protectionists at the College have obtained promising results from an integrated method that combines pruning, scaly bark removal, bait-pail and blacklight trapping, as well as clearing away deadwood, fallen leaves, apple mummies and other crop residues. The Chemistry Department in Zhongshan University is cooperating with the provincial Guangdong Institute of Entomology and Shunde County Hormone Factory to identify and synthesize the pheromone of the paddy borer – the same goals at the Shanghai Institute of Entomology. Also already considered above are some of the work on biological approaches to insect control by the Entomology group under the Biology Department of Zhongshan University.

4.3 Agrotechnology

In this section we shall look at technologies concerning chemical fertilizers and irrigation. Agriculture in China is labour-intensive and mechanization, not to be equated with modernization, has relatively low priority. On the other hand, irrigation has been applied to an extent unequalled elsewhere. In the countryside over half of all cropland is irrigated; for comparison the proportion in the USA reaches little over 15 per cent. At the same time, fertilizers dominate the agricultural chemicals sector in China: as said in the last section, fungicides, insecticides and especially herbicides are used in lesser quantities than in the West. Thus, in 1980 the output of fertilizers was 10 296m yuan in monetary terms while that of all pesticides combined together was 1 670m yuan. Chinese fertilizers consumption is so great that, despite a consistent rise in home production volume during the past three decades, making the country internationally the third largest manufacturer of urea (UNFAO figures) for example, she is at the same time still the biggest net importer of all nitrogenous fertilizers taken together. In 1980 1.5 Mt were purchased from abroad, representing about a third of the domestic demand, or 12 per cent of the total trade that year in the world market among all nations.

It should be conceded that, although her aggregate consumption is high, China maintains a level of unit utilization of fertilizers (75–85 kg per hectare) that exceeds only slightly the world average. Patient practice such as manual placement of mudballs containing fertilizers at plant roots in the paddy field serves to ensure a more efficient use of the chemicals than in many other countries. An even more influential factor is the general preference for organic farming based on enviromentally 'soft' technology. Natural fertilizers, besides being ecologically more acceptable, have the added advantage in that they can be locally produced without reduced efficiency, representing a rational implementation of the principle of self-reliance. Many forms of them are in use. Where available oil cakes as well as silt dredged from ponds and waterways are put to use. Almost everywhere composting of crop 'wastes' is done, and animal manure and night soil are assiduously collected; sometimes these substances are fermented for biogas (Chapter 7.6) before being returned to the soil. In the northern part of the country it is common to grow sesbania and clover during harsh seasons for later application as 'green manure'. At some places in the south blue algae are cultivated for the sweetening of rice paddies. It is an admirable characteristic of Chinese agriculture that together these sources probably furnish half the nutrients applied to tillage.

The situation at Shunde is a good concrete example of traditional organic farming in practice. At this county of Guangdong, sugar cane and mulberry trees are planted around ponds to faciliate supplementary feeding of fish on agricultural wastes. The fibrous vegetable matter in the cane residue is fermented into feed rich in protein and starch, whereas chrysalises and droppings of silkworms living on picked mulberry leaves are directly dispatched to the water. Ichtylic species raised are: grass carps, which inhabit near the surface; big-head and silver carps, in the middle reaches; and dace and common carps, at the pond bottom. The silt, enriched by the decomposition of their excrement, is dredged up once a year or so and distributed over the surrounding soil. The sugar cane and mulberry trees benefit also from the prevention of waterlogging by drainage channels fanning out the ponds. The whole recycling process keeps the ecology in biochemical equilibrium without the need for a significant input of artificial fertilizers. This integrated management of soil and water resources, practised for centuries by the peasants at Shunde, is now under study by scientists from the Guanzhou Institute of Geography.

Let us now turn to the production of inorganic fertilizers. The majority of the large urea plants in the country have been imported. Their feedstock is typically anhydrous ammonia synthesized from natural gas. One such complex, to supply 0.3 Mt of ammonia and 0.52 Mt of urea per year, is near completion not far from Urumqi in Xinjiang Region. In the neighbouring Qinghai province is Qarham Salt Lake, which contains a KC1 deposit amounting to 153 Mt or 97 per cent of the known national reserves. On its shore the ground has just been broken for a potash fertilizer work with a designed annual capacity of 1 Mt. The prospective leap in potash output will help to improve the 'nitrogen, phosphorous, potassium' consumption ratio in China towards the optium ratio of 6:3:2; the present figures stand at

12:6:1. Sizable, though seldom top-grade, deposits of phosphate rock and apatite exist in all the southern provinces as well as in Shandong. About 14 Mt of phosphate fertilizers are manufactured each year, three-quarters of them coming from small factories.

Local enterprises likewise are responsible for almost all the output of ammonium bicarbonate, sulphate and chloride. Although these compounds have relatively low nitrogenous contents, namely 17, 21 and 25 per cent N respectively as compared to 46 per cent N in urea, and are therefore seldom used as fertilizers in the West, together they make up over a third of the annual production of nitrogenous fertilizers, in China, with the bicarbonate being the most important. Their lower fertilizer values are compensated by the generally shorter distances over which they have to be shipped. Moreover, the gestation period of a large-scale plant may exceed three years whereas the lead times for small factories average one year. Many of the decentralized factories date from the period of the Big Leap Forward, when the policy of 'walking on two legs' and the spirit of self-reliance at the commune or even brigade level were first stressed. Today, some serial manufacturing of equipments for such installations still goes on in Shanghai, Tianjin and Dalian. A typical factory takes lignite or coke dust as the feedstock, which is screened, sized, and then briquetted before being brought into reaction with steam in a reformer. After the removal of methan (for making alcohol) and oxygen, carbon dioxide and monoxide are separated out under high pressure, and then the syngas is passed over catalysts. The ammonia so synthesized is then dissolved in water. Some of the ammonium hydroxide may be shipped in tanks to farm fields, but usually most of it is reacted with the recovered carbon dioxide into bicarbonate solution, which, upon conversion into solid granular form, is easier to transport in China, where pipeline distribution has not caught on.

Some organizations in the Academy of Sciences system are active in research related to fertilizers. Concerning fertilizer application, the Nanjing Institute of Soil Science, an establishment formally founded in January 1953, has for instance made contributions to the rational exploitation of ammonium bicarbonate. A method was evolved for injecting the chemical deep underground to reduce its volatility and thus raise its utilization index. Processes were also invented for turning it into pressed pellets without the addition of glutinating agents, and for coating the pellets with a calcium-magnesium-phosphate substance in phosphoric acid to induce long-acting properties. Regarding fertilizer production, both the Dalian and the Lanzhou Institutes of Chemical Physics (of these two more later in Chapter 7.2) have discovered catalysts for ammonia synthesis. Rare earths, of which China possesses unsurpassed geological reserves as will be considered in Chapter 6.5, are used not only as methanation catalysts but also directly as additives to fertilizers. The Qinghai Institute of Saline Lakes, to be more fully described in the next chapter on pp 105–6, is a major centre for studying the extraction, from salt lakes and their sediments, of potassium, iodine and other compounds essential to agriculture. The Northwest Institute of Soil and Water Conservation (of which more below) has worked out a method of using, as a fertilizer, underground water into which significant amounts of nitrates have been dissolved.

The Ministry of Chemical Industry directs all large-scale fertilizer plants. One of the research establishments under its central administration is the Shanghai Chemical Industry Research Institute. The Institute was instrumental in many major advances in the design of fertilizer complexes for manufacturing phosphoric acid, and ammonium, diammonium, triple and super phosphates. The electricity generating industry has also made contributions. In 1976 the Wuhan Power Station successfully tested a technique of burning phosphate rock in vertical cyclone furnaces to generate power and producing fertilizers from the ashes and dregs. This innovation represents a step forward not least in pollution control.

The investigation of organic farming by the Guanzhou Institute of Geography has been mentioned above. As a further example of work on natural fertilizers, the cultivation of duckweed as a source of green manure (and animal feed) was introduced by the Institute of Soil and Fertilizers subordinate to the Zhejiang Academy of Agricultural Sciences. Efforts are being continued at this Hangzhou Institute to explore the utilization of alternative forage grass as fertilizers (and fodders). Among the other provincial institutions, the Agricultural Test Centre under the Hebei Academy deserves special reference as it is the first of nine newly constructed test centres to open, in late 1982. Projected ultimately to handle half a million soil samples a year, it has up-to-date analytical instruments and electron microscopy systems, many of which incorporate slave computers for online data reduction. Its Inorganic Fertilizer Analysis Laboratory specializes in red-yellow soil (ie laterite and ultisol), an acidic type of soil common throughout southern China.

We now turn to farm irrigation, a feature prominent both spatially and temporally. As said in the beginning of this section, over half of China's farmland is irrigated, ie, provided with water through permanent installations such as ditches, catchment basins, dams, wells and pumps. Historically, the Chinese built a large-scale irrigation system over twenty centuries ago; called the Dujiangyan, it is still functional today and helps to enhance the fertility of Sichuan's Red Basin. More recently, during the Cultural Revolution, the Red Flag Canal was completed in northern Honan and thence raised the area of stable high-yield cropland from 800 to 40 000 ha. It has been argued by a Western observer in 1978 that 'improvements in water control are the most important single development in Chinese agriculture in the past twenty-five years.'

Indeed, water conservancy projects receive both high priority from the government: besides offering the obvious benefits to farming, they promote the socialist sector in the small-peasant economy because of their collectivistic nature. In the First Five-Year Plan (Appendix 2) they constituted the only major item on the state budget for agriculture. The Great Leap Forward was originally conceived as a way of mobilizing the peasant masses to undertake them. The Draft Plan for Science and Technology (1978–85) assigned priorities to the studies of 'science and technology for improving soil, controlling water . . . and projects for diverting water from the south to the north.' The project has now actually already been started for diverting water from the Yangtze, to the North China Plain which has 51 per cent of the country's cultivated area but only 7 per cent of the surface water flow, compared to 33 and 76 per cent respectively in the case of the Yangtze river basin. The first stage, to be completed by 1990, involves the reconstruction of the middle section of the 1 000-year old Grand Canal, between Hangzhou and Jining, a city just south of the Yellow River.

Due to erratic rainfall and insufficient drainage, the North China Plain is vulnerable to drought and waterlogging at one time or another. The ambitious project just described is designed to tackle the problem once for all. In the past, partial solutions have been sought through a combination of engineering work on the Huai, Yellow and Hai rivers and tributaries, as well as the exploitation of artesian water. Near to three million wells have been sunk, varying in depth from 30 m to as much as 250 m, and about a tenth of which are power-operated. However, the aquifer in the Plain being bedrock, the drawing of water needs to be moderated, if serious decline of the water table and accompanying subsidence are to be avoided (as are soil salinization and alkalization, where drainage is slow but evaporation fast). The government is spending part of the US$60m loan taken out in 1982 from the World Bank on digging drainage canals, building pumping stations and levelling farmland, the work being expected to be finished by 1987.

In terms of installed power ratings the mechanization of China's agriculture was put in the 1982 CIA estimates at 210m hp, of which roughly half are accounted by water moving equipments. Little over a tenth of arable land is ploughed with tractors, still lower fractions are sown and reaped with powered mechanical devices, but tractors are often employed in the role of rural transportation. As far as pumps are concerned, a fifth of their capacity is engaged in irrigation and the remainder in drainage, according to an American Water Resources Delegation on a recent visit. Steam, diesel, gas and electric pumps are all in use; very infrequently a traveller may come across kerosene burners also. The number of electric pumps is gaining in proportion to those of the other categories. There is a wise preference to conserve diesel, essential for tractors, but to rely instead on electricity, the availability of which has resulted from the remarkable success in rural electrification, now standing at almost 100 per cent, and has been enhanced by the governmental policy of charging electricity less in villages than in cities. Rural electricity is mostly consumed for water conveyance purpose: less than a third is used by other farming machineries, in the household, and by light industries in the countryside. Fittingly, much of it is supplied by local hydropower stations that are themselves often integrated with irrigation and flood control systems (Chapter 7.3).

A great variety of small electric pumps are produced, some of which have been exported. Types of both mixed and axial flow, with either horizontal or inclined axia, are made. Of the bigger kinds, an example is the vertical centrifugal model, manufactured in Shanghai, that has a 3.1 m-diameter calibre and can displace 30 cubic metres of pumpant a second by the action of a 3 000kW synchronous motor. In the typical irrigation system, at the pump outlet water is allowed to flow under gravity through dug channels on the ground and whence diffuse into the soil, the optimum spacings and depths of these channels being dependent on the local conditions. At places this traditional arrangement may be modified by the installation of plastic irrigators. One case of recent technology transfer from abroad is the introduction in 1980 of laterally mobile sprinklers designed by Lindsay International of Columbus, USA

Basic research on farm machineries goes on in the Institutes of Agricultural Modernization at Changsha, Shijiazhuang and Harbin. These three organizations under the Academy of Sciences were all set up in 1978, when the Academy witnessed a spectacular expansion by regaining certain research centres that had been transferred to particular ministries, reviving others and establishing new ones. The Changsha Institute comprises six divisions, specializing in, apart from agricultural engineering and applications of novel technology, also resources utilization, physiochemical analysis, agricultural ecology and agricultural economy. Field experiments are carried out in the nearby Taoyuan county. The Shijiazhuang Institute has five divisions, with the respective titles of Farm Machinery, New Technology, Resource Survey, Agricultural Ecology and Agricultural Economics. Its experimental base is Luancheng county. The Heilongjiang Institute, located at Harbin and linked to Hailun county, is divided into six parts: agricultural machinery and agronomy, reserves exploitation, comprehensive testing, energy sources, ecology and agricultural economy. The Northwest Institute of Soil and Water Conservation has existed at Wugong in Shaanxi province since 1956. Earlier has been mentioned its work on groundwater as fertilizer. One of its nine divisions is concerned with the engineering aspects of its title subject, although until some years ago its activities tended to bear exclusively on the surveying of soil and water resources in the middle reaches of the Yellow River. Its Soil Erosion and Soil Geography Divisions will be discussed on p 101.

Under the Academy of Agricultural Sciences, the Agricultural Mechanization Research Institute designs tractors, ploughs, seeders, transplanters, threshers, combine harvesters and the like, but not pumps. An interesting point to note, however, is that several multipurpose tractors from its drawing board (and in production) include as one of their functions the opening of irrigation channels.

The government body responsible for irrigation work used to be called the Ministry of Water Conservancy, which had in 1958 absorbed the Farmland Irrigation and the Irrigation Control Bureau originally subordinate to the Ministry of Agriculture. Any river taming projects that entail the construction of dams taller than a specific height were conventionally put under the charge of the then independent Ministry of Hydroelectric Power. The two ministries have since been merged, separated, and recently merged again into the present Ministry of Water Conservancy and Power. The largest of the central laboratory run by it is the Water Conservancy and Hydroelectric Power Research Institute in Beijing. Pumps as well as sprayers, control devices, meters etc are designed and trial-produced by staff in the division for hydraulic and electric machinery of the Institute. Activities of some other divisions will be sketched in Chapter 3 of the next chapter and in Chapter 7.3.

The Ministry excercises authority with the Education Ministry over several institutions of tertiary education. Among them, the Wuhan Hydro-electric Power College, the Hebei Electric Power College at Shijiazhuang and the East China Water Conservancy College in Shanghai have been named as key universities. Work on irrigation equipment may also be found in certain polytechnic and agricultural colleges. The Water Conservancy Department in Chengdu Polytechnical College, for example, has four fields of specialization: farmland irrigation, irrigation engineering, hydrology and hydroelectric power. Two departments in the Northwest College of Agriculture at Wugong are in agricultural engineering and hydrology; a third is in forestry.

4.4 Forestry and Desert Control

The eastern half of China is suited for forest growth, having annual precipitation of 400 to 2 000 mm and sometimes more. However, vast domains that, according to historical records, once supported a rich forest flora have shrunk thanks to continous encroachment by shifting cultivation, overgrazing and indiscriminate felling. Probably as much as one-third of the national territory is land that was farmed piecemeal at various times but had since been abandoned and left to erode into deep ravines. Although as a result of vigorous efforts by the government 30 m hectares have been reforested or afforested since 1949, the aggregate forest coverage is still a mere 122 m ha, which translates to 12.5 per cent of the national land area, half the average for countries of the world, or to 0.12 ha per capita, compared to 0.32 and 3.4 ha per person in Western Europe and the USA respectively.

However, surveys have characterized an additional 140 m ha as prospective forestland. The government's goal is to realize the potential in half of this area by the end of the century. 'Green the Country' committees have been set up at both provincial and county levels all over the country. The Sixth Meeting of the Standing Committee of the Fifth National People's Congress approved in February 1979 a provisional forest law, and also decreed that the 12th of March be recognized as the National Tree Planting Day. Subsequently, every able-bodied citizen over the age of eleven has been called upon to plant and care for five trees annually, in a resolution passed by the Fourth Session of the Fifth Congress in December 1981. In agricultural regions of the south, 'four sides' (referring to waterside, hillside, roadside and homeside) plantings are being pursued en masse, in earnest. In the north, a gigantic drive commenced in 1979 to build a 7 000 km long shelterbelt from Xinjiang eastward to Heilongjiang. The aim is that 6 m ha of woods cover will be created by the end of the first phase of this 'Green Great Wall' project, in 1985 at the close of the Sixth Five-Year Plan (cf. Appendix 5). In addition to these constructive efforts a strict legal code is in force to prevent irreversible destruction of forestland. The Ministry of Coal

Industry is made responsible for implementing reforestation programmes at all state-owned coal mines. Upon the completion of major railway and road constructions, the Ministries of Railways and Communication are required to 'green' relevant areas. If all these measures prove effective, China will literally change her appearance in not many years' time.

In terms of utilization, existing natural forests and plantations may be delineated by areas as follows: 80 per cent timber resources, 6 per cent shelterbelts, 7 per cent miscellaneous products (tung oil, lacquer, camphor, pecans, chestnuts, tea oil, etc), 3 per cent bamboos (Phyllostrachys spp.) and 4 per cent others. In terms of vegetation taxonomy the division is complex and warrants a careful approach to planning and research, especially in silviculture. An incomplete inventory tallies about 3 000 species of trees; by comparison, the number for the USA has been put at 679. New species and even new genera are still discovered from time to time, particularly at remote mountainous spots. Some 30 genera of Pinaceae and Taxodiaceae have been counted in the world; China nurtures 20, among which Cathaya, Taiwania and six others cannot be found elsewhere. Since the mid 1950s a great many forest surveys of different scales and with various objectives have been conducted. The major ones often involve silviculturists, dentrologists, soil scientists, botanists and ecologists working in teams. Some were concurrent with activities of even broader scope, as parts of comprehensive land resource surveys of specific regions. Tree specimens, soil profiles, and specimens of arborous pathogens and insects are collected and mostly identified; volume and yield tables were also prepared. These results provide a firm foundation on which further specialist research can proceed.

Increasing importance is being attached to tree improvement programmes. Stocks for roadside plantings of fast-growing species such as Populus for the north and Platanus in the south are obtained, often through cloning, from tree nurseries found in the outskirts of many cities. However, there is now less insistence on genetic uniformity, with the realization that genetic variety also needs to be attained if viability is to be maintained. Tree selection for resistance, growth rate etc, started in the early 1960s and by now over 80 000 strains have been selected. Genera represented include Populus, Pinus, Cunninghamia, Paulownia and Eucalyptus. Provenance tests were begun in the mid 1960s. Several generations have been examined for timber trees such as Pinus massoniana, P. tabulaeformis, Cunninghamia lanceolata and Paulownia fortunei, tomentosa and elongata.

The work on genetic breeding is gaining new impetus from the progress in anther culturing techniques, which were recently applied to Populus, Cunninghamia lanceolata and Paulownia spp. Efforts along the more conventional direction of hybridization have already yielded invaluable results. Populus is the genus most studied. The majority of the popular hybrids are crossed of native species (cathayana, simonii and others) with introduced cultivars (nigra and its varieties, canadensis, etc) and grow faster than their parents. Success has also been met in hybridization between genera. As an example, Cunninghamia lanceolata was crossed with Cryptomeria fortunei and with Taxodium distichichum.

Tree introduction is a burgeoning activity, especially in the south. Promising species to date include Khaya senegalensis, Swietenia macrophylla, Acacia auricumiformis and mearisii, as well as several eucalypts. Exotic trees are being tested at a number of institutes and nurseries. Among those that have been widely propagated are Populus canadensis, Taxodium distichum, T. ascendens, Larix leptolepis and Robina pseudoacacia.

Pest control has made steady progress. The insect responsible for the greatest damage is probably Dendrolimus punctatus, which annually defoliates large areas of Pinus punctatus plantations. Another harmful pine caterpillar is D. spectabilis on Pinus tabulaeformis. Other major problems concern wood borers, leaf-casting beetles, leaf caterpillars on poplars, budworms on pines, the bamboo locust and Polychrosis Cunninghamicola on C. laneolata. The biological approach is emphasized in the integrated management of these insects. As in the case of grain and vegetable crops, Bacillus thuringiensis, Beauveria bassinana and Trichogramma spp. are among the insecticides commonly applied. In the northeastern city of Changchun, subordinate to the Academy of Sciences, the Institute of Applied Chemistry, of which more on pp 134, 136 and 140 has characterized and synthesized pheromones from several female pine caterpillars.

The protection, selection, breeding, introduction and acclimatization of tropical and sub-tropical trees constitute one of the main preoccupations at the South China Institute of Botany, which we have already met in section 1 of this chapter. This institute at Guangzhou is in charge of the Dinghu Mountain arboretum that extends over 1 127 hectares, encompasses an 167-ha forest with a history of over 4 000 years, and contains the nationally unique collection of bamboo with more than 100 species. The arboretum, together with Changbai Mountain and Wolong District reserves (the latter famous for its giant pandas) in Jilin and Sichuan respectively, have been listed since 1980 by UNESCO in the World Network of Nature Conservation, and are subjects for the UN's MAB study programme. The Yunnan Institute of Tropical Botany was founded in 1959 and is situated in Xishuangbanna, a prefecture noted for its scenery and containing many virgin tropical forests – but in parts threatened by desertification, as will be explained later. One of the institute's divisions carries the title 'Plant Introduction and Acclimatization'; the others are Plant Classification, Experimental Plant Community, Plant Physiology, and Phytochemistry. Satisfactory outcomes have been secured with such scarce trees as Anthocephals chinensis, Parashorea chinensis, Ochroma lagopus, Gmelina arborea and Dalbergia fusca. Back to the northeast, at Shenyang, there is an institute for forestry and soil science. It began in 1954 and now runs seven departments. The one for forestry has contributed to the subjects of forest protection, tree growth, felling and renewal of woods, shelterbelt for farmlands, and sand fixation by afforestation.

State affairs in forestry matters are overseen by an exclusive government ministry, and relevant research promoted by an Academy of Forest Science funded by this Ministry of Forestry. In addition, research centres have been set up on the provincial level. There is, for example, one at Changsha in Hunan, which is conducting pilot tests of different approaches to the biological control of Dendrolimus punctatus. The control of this caterpillar with the help of Trichogramma has also been looked into at the Jilin Institute of Forestry Research at Changchun. A series of publications has originated from the Shaanxi Institute at Xian, concerned with Gravitarmata margarotana which attacks pine trees in early summer. The Hubei Institute at Wugong is an authority on Taxodium ascendens, introduced from North America and widely planted around the middle and lower reaches of the Yangtze. One of the special interests of the Zhejiang Institute is the cultivation of camellia forest for its oil-producing capacity. The Yunnan Institute at Kunming has made detailed studies of the production cycle, propagation, walnut yield and oil yield of local walnut trees.

The country has eleven forestry colleges, as well as sixteen agricultural colleges, of which the Northwest College of Agriculture mentioned in section 2 is an example, that have forestry faculties. Five of the forestry colleges come under the sole jurisdiction of the Ministry of Forestry, but the rest, which tend to be bigger institutions, are jointly controlled by the Ministries of Forestry and Education. Designated by the regions served, these six are respectively at Harbin (northeast), Beijing (north), Xian (northwest), Nanjing (east), Changsha (south central) and Kuming (southwest). The last has been named as a key university. As an illustration of research carried out in these teaching institutions, work on the taxonomy of willow trees has a strong presence at the Northeast College of Forestry.

The admission has been made that up to perhaps a third of China's land mass was forested area now denuded. Each year deforestation causes an estimated 5 000 m tonnes of soil to be washed away, accompanied by a loss in nitrogen, phosphorus and potassium that exceeds the entire output of the Chinese chemical fertilizer industry. The erosion, by silting up lakes and raising riverbeds, is also partly responsible for recurrent floods. Furthermore, it is thought that since 1949, the state's effort on reafforestation notwithstanding, damage to windbreaks and shelterbelts have let 6 m ha of land turn into desert. Indeed, the proportion of desert area in the country is showing a net increase: it was 11 per cent in 1977 but now stands at 13 per cent, equal to the fraction of cropland. Airborne sand in Beijing, noticeable especially during winter when there is a strong northwesterly wind, has become a greater health threat due to the southward expansion of deserts in Inner Mongolia. In the south, parts of the Pearl River valley and Hainan island, both of Guangdong province, and Xishuangbanna in Yunnan are in serious danger of degradation into sandy lands.

Given the weight of the problem, it is no surprise to learn that desertification receives close attention both in practice and in research. A number of communes have set up desert control teams, and the Academy of Sciences administers an institute that specializes in desertology. Investigations conducted at this institute into desert distribution and types indicate that just over half of China's desert area is of the dune type, the rest being either gravel plains or sandy sheets. Gurbantunggut, China's second largest desert, is covered by gravels and sand dunes (15–25 m high and mostly fixed) in about even proportion. Three-quarters of Taklimakan, the biggest, consists of moving dunes up to 200 m in height.

The Chinese have rich experience in preventing or even reversing desertification. A common measure is to plant wind-resistant shrubs across the main wind direction around deserts and tall trees near farmland. For such a purpose poplars are popular, having deep roots and fast growth rates. Capparis, a semi-shrub rattan that yields oil-bearing seeds is also a common choice. At moderate wind speed the length of protected area may be over twenty times the depth of the forest belt. The 'Green Great Wall' project referred to earlier serves as a grand example of this preventive method. A similar approach has been adopted in the reclamation of dune deserts. Shrubs are planted on the lower third of a dune's windward side, to lower the wind speed near the bottom of the dune which is thus anchored. Unaffected, the wind at the top of the dune level it off. Trees are then grown on the flattened dune. In this way the vegetation cover can be increased by 50–80 per cent after five years.

Another technique, followed in many terrains with fixed sand dunes or only interdispersed shifting sand, involves developing pasture land for animal husbandry. Where groundwater resources are available, forage bases are created on flat land in between the dunes. Such 'kuluns' may be established to prevent overgrazing of nearby areas by providing fodder, or to control sand movement. Desert marshes, if present, are upgraded by digging canals and by planting belts of tall trees perpendicular to the wind direction. Forest strips spaced 50–100 m apart have been found to be most effective.

Oases on the fringes of deserts are protected from desertification by the construction of shelterbelts 50–60 m wide along their perimeters. Within several years, the wind will blow 1 to 2 m of sand off the tops of adjacent dunes, hence reshaping the relief of the land. Planting vegetation at this point will improve the quality of the surface material and, since the water table is usually high near oases, the former dunes can be transformed into productive farmland eventually. The Chinese, furthermore, create oases by building reservoirs to hold seasonal flood water from streams, and use melted snow from the mountain or groundwater to irrigate the new farmland before planting trees to consolidate the reclamation. In a windy gobi region, an oasis is best protected if it is rimmed by specially selected grass and shrubs. An inner forest belt stop any sand that manage to pass through. In addition, small plots of trees are scattered inside to help stabilize the area.

A research base for desertology is the Xinjiang Institute of Biology, Pedology and Psammology in the Academy of Sciences. Founded in 1961, it operates the Turpan Sand Control and the Mosowan Desert Control Stations, where wind protection, sand fixation and desert ecology are studied. Its other main research items concern the regional distribution and utilization of grassland, soil, wild plants and bacteria: the Youlds Grassland and the Xinhe Soil Reforming Experimental Station also belong to it.

Another base is the Lanzhou Institute of Desert. In 1958 the Academy organized a Desert Research Group in Beijing. Seven years later the Group was moved to Lanzhou and merged with the Laboratory of Glaciology and Cryopedology under the Institute of Geography, to form the Lanzhou Institute of Glaciology Cryopedology and Desert Research. Finally in 1978 the desertology part split off again and became a full institute on its own. It currently has active programmes on aeolian-sand movement, processes leading to desert formation, trends of desertification, the distribution and characteristics of China's deserts, their natural conditions and their resources. It also directs its energy towards sand control via biological and chemical approaches, investigating ways of combating sandstorms, the utilization of soil and water resources in sand areas and reclaimed land, and preventing sand damage to rail lines. The last aspect of work is dealt with at its out-station at Shapotou on the southeastern corner of the Tengger Desert. More than 40 km of the rail link between Yinchuan and Langzhou passes through that corner region, where dunes move outward by as much as 15 m a year. Protective measures evolved at Shapotou as well as by the desert research department of the Academy of Railway Sciences will be described in Chapter 8.1.

4.5 Environmental Science

It has been a traditional Chinese mode of thought, now reinforced by modern conviction, that mankind does not stand apart from the 'secular' world but rather, in blood and flesh, exists within nature. Systematically defined responsibilities and limitations regarding the internal relations within the whole of man and nature were laid down by the first century BC. Such awareness was reflected in, for example, the genius for water and wastes management displayed by the ancient Chinese. However, owing to various historical factors at one time or another ecological abuses could not altogether be avoided. In particular, during the past three decades serious pollution problems have arisen out of the single-mindedness in expanding the national economy. For instance, the smog in Beijing was reportedly enough to overload the recording instrument of a visiting UN environmental team. The frightful fact was disclosed that, of the 2 400 localities throughout the country where tests were conducted by the end of 1982, nearly 45 per cent suffered from acid rain. Nevertheless, the year 1973, when the State Council of the central government held its first meeting on environmental protection, may be seen to mark a more active phase of ecological conservation. A direct result of the meeting was the establishment of the Environmental Protection Office, which will hereafter be referred to as the EPO. Later, in September 1978, in response to UN initiatives, a 'Man and Biosphere' National Committee was also set up.

The EPO is an administrative unit directly responsible to the State Council; there are three other 'Offices': for Air Defence, Birth Planning and Overseas Chinese Affairs respectively. Offices have lower status compared to government bodies designated as 'specialist agencies' or, on a still higher echelon, as 'commissions'. In the case of EPO its subordination to economic considerations is reflected by its location within the State Capital Construction Commission building in Beijing. It is imbued with the work of formulating and implementing relevant policies, legislation, coordinating activities over the whole country, and disseminating information through lower levels including popularizing among the man in the street.

China's first environmental protection law was passed at the Eleventh Meeting of the Standing Committee of the Fifth People's Congress convened in September 1979. Chapters Three and Six of this bill, which has provisional status, specify the standards required of the industry in reducing pollution and the measures for their enforcement. Accompanying the promulgation of the law, 167 organizations were directed by the State Economic Commission to clean up by 1982; subsequent to this deadline several enterprises have been fined. Margaret Trudeau has recalled how, while on a boat ride on the picturesque River Li near Guilin during an official visit in 1973, she communicated to the host her horror in finding a serious polluter in the form of a soap factory at the foot of the mountain. According to a Chinese press account in 1982, during the previous five years at that scenic spot a foundry and the pulp workshop of a paper mill were shut, and a printing house, a dyeing plant and an electroplating installation were relocated.

Chapter Five of the provisional law consists of two paragraphs that underpin, separately, the need for continued efforts in popularization and research. It has been fully realized that the involvement of the masses is an essential component to any successful ecological campaign. In 1980 a month was devoted to environmental protection awareness, an event resembling the Earth Days in the USA. For research, the EPO maintains a science and technology division the functions of which are to coordinate, monitor and assess programmes conducted at different institutes.

Environmental science has been recognized as an independent discipline in China only since the early 1970s, but it got on the list of the 27 priority spheres of research in the original 1978–85 Science Plan. Already, during the past decade much has been achieved. Surveys on regional pollution were completed for Shenyang, Nanjing, Shanghai and Maoming, apart from the western and the southeastern district of Beijing. Among the waterways, Di'er Songhua, Tumen, Huangbao and several other smaller rivers, sections along the Grand Canal, as well as Lake Ya'er near Wuhan have been charted, and so have Bo Hai, the northern part of the Yellow Sea, the East Sea and the South Sea. These surveys covered the identification of main pollutants and their sources, their quantification, and the examination of related geographical factors. On the basis of these data, environmental quality indices and environmental capacities have been evaluated, and some variant analyses and modelling of individual pollutant mobilities carried out. The background concentrations of specific chemicals have also been measured in many areas. For example, those of heavy metals in the soil are known for all the suburbs of Shanghai. Atmospheric pollution used to receive less attention but, in July 1979, the first national conference was held on this subject at Shenyang. The occurrence, in some industrial cities, of photochemical smogs that were not caused by emissions from car exhausts was confirmed at the meeting.

A series of investigations have been conducted into the effects of heavy metals and chemical pesticides on the paddy field and the inland water ecosystem. Substantial results were obtained concerning the accumulation, transportation, metabolism and breakdown of these pollutants in biological organisms. A spectrum of plants that are resistant to sulphur dioxide, hydrogen fluoride and chlorine gases have been selected. Major advances were made in applied pollution chemistry. For sulphur dioxide at low concentrations, a highly effective method of removal was evolved which relies on the absorption by activated carbon containing iodine. In addition, the phytopathological mechanism of this dioxide has been clarified. It was shown that the cell plasma membrane is easily damaged by the dioxide, leading to the outward diffusion of potassium ions and consequently the over-oxidation of membrane phospholipid. Unlike the case of many other countries, oil refineries in China are not much to be blamed for sulphur oxide pollution. This happy situation results from the commonly low (meaning <0.1 per cent w/w) contents of sulphur in Chinese crude – the sole major exception being Shengli crude, which may yield up to 1 per cent – and also to some extent from a considerable consistency in implementing integrated programmes of pollution control and wastes recycling. Daqing refineries, for instance, have successfully utilized a series of processes for recovering sulphur from effluents, such as oxidation, the use of copper chloride as a catalyst, and steam stripping at high temperature and pressure whereby hydrogen sulphide and ammonia can be recovered. On the other hand, sulphur dioxide coming out of the venting by coal-burners constitutes a big problem. It has been said earlier that acid rain is a serious threat to the country.

The Dalian Institute of Chemical Physics has designed for use by the Daqing refineries purifiers that reduce the emission of ozone and nitrogen oxides. To deal with phenolcyanide wastewater from coking plants, the Institute pioneered an efficient procedure, whereby coal gas is employed to blow out cyanide which is then burnt in hot-blast stove. A similar technique has been devised by the Institute for wastewater from dyestuff processing that releases nitrophenol and nitrochlorobenzene. Chlorobenzene recycling is done to enhance the absorption by charcoal, solving the need for carbon re-activation, and producing chemical by-products as an added benefit.

The wheat crop in several regions of China was once adversely affected by industrial trichloroacetic aldehyde (choral). On the basis of field observations and laboratory experiments, a theory was formulated for the underlying causal links, the most crucial of which was the oxidation of choral to trichloroacetic acid due to actions by microbes in the soil. Preventive and remedial measures based on the theory have since be taken and the wheat harvest restored. Another troublesome substance is DDT. Although its use has now been replaced as far as possible by biological and cultural methods of pest control (Chapter 2), owing to its persistence DDT remains a major toxin in fresh water resources. The trace analysis of it along with the heavy metals is routine at many research service centres. Its analysis is typically performed with the help of gas chromatographs, versatile models of which are manufactured by the Chinese in adequate quantities. Direct polarography, anodic stripping voltametry and atomic absorption are the usual means of determining heavy metals in preconcentrated samples. Atomic absorption instruments of both home and foreign origin, and with flame or graphite furnaces, can all be encountered. A recent innovation in the treatment of DDT is catalytic reduction with two-metal systems.

The country places equal emphasis on ecological management as on passive remedies to pollution. As far as possible environmental planning and economic productivity are viewed as objectives that go hand in hand. The use of biomass as natural fertilizers, non-chemical approaches to pest control and preventive actions against desertification, aspects which have been examined in the last three sections, may be cited as illustrative of Chinese practice in environmental management.

With regard to environmental science, the most conspicuous research establishment under the Academy of Sciences is the Institute of Environmental Chemistry. The Institute came into being in 1975, with its staff mostly drawn from the Institute of Chemistry, as a response to the call for greater attention to ecological conservation by the government in 1973. It now has five divisions, for: Inorganic Analytical Chemistry, Organic Analytical Chemistry, Pollution Chemistry, Pollution Control and Treatment, and Instrumentation. In addition a Food Synthesis laboratory is attached. The technique for catalytic reduction of DDT mentioned earlier has been proposed by it. Examples of its other contributions are the treatment of wastewater contaminated by Sumithion (a pesticide) via solvent extraction, removal of nitrogen oxides by absorption in dilute hydrochloric acid, and closed-circuit chrome plating with no discharge. Apparatus designed includes a cyanide ion detector, a CoD meter, and an ozone monitor for atmospheric surveillance.

Analytical instruments for measuring pollutants in solution used to be more readily available than those for air pollution, but the latter have become more and more common. A commonplace example is the Zhp-3 quadrupole gas analyser supplied by the Academy's Beijing Scientific Instrument Factory. This simple but robust spectrometer has a mass resolution of 160 at 10 per cent valley, and a minimum detectable partial pressure of 0.3 nPa. Earlier in 1983 China placed an order with Vacuum Generators Ltd. for five ZAB-3F magnetic mass spectrometric systems, and it is understood that one of them will be allocated for environmental quality control applications.

One of the fourteen divisions in the Nanjing Institute of Soil Science, also under the Academy, is entitled Soil Environmental Protection, being concerned with the pollution of soil and crops by heavy metals and pesticides. Likewise the Institute of Geography, an overall description of which can be found in section 3 of the following chapter, has formed in 1974 an environmental protection group. This group is studying the migration and control of toxic chemicals in water bodies. Work is also in progress aimed at health promotion through a greater understanding of the effects of groundwater composition.

The Shangdong College of Oceanography, of which more in section 4 of the next chapter, has conducted a number of oil pollution surveys in the Yellow Sea's coastal waters. The work has not been extended to the research into biological effects of the hydrocarbon pollutants. On the other hand, in the context of freshwater resources, the Institute of Hydrobiology at Wuhan has contributed much to 'water pollution biology' – which serves as the title of its latest division to be established. (There are five other, older divisions: see section 2 of the following chapter.) The scope of the work at this newest division encompasses ecology, environmental toxicology and the amelioration of polluted water. Experience has been gained in the use of three oxidation ponds in series to treat pesticide wastewater, and of plant materials to breakdown chlorinated hydrocarbon insecticides from agricultural runoff. The feasibility has also been proved that, with the help of Phomidium ambiguum, a blue-green alga, hydrogen can be taken out of vinylcyanide 'waste' water and turned into ammonia for utilization as a fertilizer. Effluent discharged from synthetic fibre plants and containing polyacrylonitrile can, on the other hand, be purified by vinylcyanide-oxidizing bacteria, as has been shown by the Institute of Microbiology. The Institute of Organic Chemistry in Shanghai has worked out methods of employing ion-exchange membranes to recover several valuable materials from industrial wastewater by electro-osmosis. It also successfully tried out the electrodialyzer removal of mercury in the production of sodium hydroxide. A solvent extraction process (with N503 as the extracting agent) was developed by the phenol content of phenol-containing sewage may be reduced to the 80 mg/litre level.

Research on environmental science in institutions of higher education has intensified after a symposium was called in 1978 by the Ministry of Education. Nevertheless, while related subjects have since been introduced into the teaching curriculum at a number of places, none of these universities has created independent departments devoted to terrestrial ecology. At Beijing Normal University, work on environmental quality evaluation is undertaken within the Department of Geography; such an arrangement is typical. Projects concerned exclusively with the analytical aspect of pollutants are, however, underway in chemistry departments frequently. Chemists in Fudan University, for instance, are noted for their expertise in anodic stripping voltametry. They have demonstrated the detection of copper, zinc, lead and cadmium ions present in wastewater at the ten parts per million million level. Lastly, a few government ministries run research centres that look at pollution problems specific to their spheres of responsibility. As an example, the Institute of Environmental Protection funded by the Ministry of Agriculture is preoccupied with those problems caused by pesticides.

5 Earth Sciences

This chapter is concerned with sciences of our home planet, beginning with its inner atmosphere. For such a large sphere to be covered in these pages selections are obviously necessary. Thus, subjects that have less direct applications, like physics of the mantle and the core, will be skipped and so will be the more descriptive disciplines, such as stratigraphy. Things which have been included in the chapter on environmental sciences will also be absent here. The scientific aspects of geophysical prospecting will be included for discussion in the next chapter. On the other hand, seismic engineering, though strictly speaking not part of earth sciences, will be covered in the section dealing with seismology and earthquake forecasting.

The people of China have a long history of interest in the large-scale study of their physical environment. In meteorology, we find for example that quite comprehensive weather records were kept as far back as the thirteenth century BC and that the wind vane was in use before AD132. Concerning the ocean, tables of coastal tides have existed since the ninth century or earlier. River-valley erosion was referred to by a poet-statesman of the second century BC and no less than 137 rivers were identified and described in the first-century work 'Book of Waterways'. In 'Ode to the East Capital', written not long after, an investigation into the chemical components of subterranean hot water was reported. The Chinese contributions to magnetism and geomagnetism need no reminder. In geology, the idea was discussed more than eight hundred years ago that the presence of conchs and oyster shells within the rocks of some mountains could be explained by the uplifting of the seabed. Systematic recording of earthquakes date from the twelfth century BC and the first seismograph was constructed in AD132.

5.1 Atmospheric Disciplines

Climatology and meteorology are given great emphasis in China because of their relevance to agricultural production. Large-scale, long-term studies of weather systems began soon after 1949, with many of them coordinated by the Commission for Integrated Survey of Natural Resources, under the Academy of Sciences. A lot of effort has been spent in determining nationwide patterns of temperature and precipitation, which formed the basis for the identification of optimum crop zones. Attention is presently focused on more fundamental investigations, such as those on climatic changes over long time scales, oceanic influences on typhoons and other tropical weather events, mechanisms of basic physical processes in the lower atmosphere and energy flow in the biosphere. Long-term climatic forecasting has been acknowledged to be a priority subject for several years now (Working Group of Atmospheric Science report, June 1977).

A topic related to climatic forecasting is palaeoclimatology, on which the Chinese are in a privileged position to work. China has the longest uninterrupted documented history of all countries. Official dynastic records run to thousands of volumes, supplemented by countless local annuals, family chronicle, diaries and other relevant literature. The phenological approach to studying climate in the past was pioneered by the late K. Z. Zhu. It has yielded a wealth of unique data as well as results that are on the whole consistent with those inferred from other sources, like oxygen-18 profiles of ice sheets, sequences of tree ring thicknesses, and histories of sea levels, water levels in lakes and snow lines on mountains. It is being continued chiefly at the Institute of Climate of the Central Meteorology Bureau (of which more below). The picture that has emerged suggests that in the last 5 000 years the annual mean temperature fluctuated within 6 Centigrade degrees and rainfall within 500 millimetres, but the trend was for both fluctuations to decrease in amplitude. Four warm periods (seventeenth to eleventh century BC, 770 to 10BC, 580 to 910, and 1206 to 1368) and five cold spells (twenty-first to twentieth century BC, 25 to 500, 1000 to 1200, and 1400 to present) are evident, the tendency being for the cold spells to extend longer and longer. Periodicities of 90, 35, 20 and 2–3 years can also be noticed in the time series of average precipitation.

Research is also being undertaken into prehistorical climates. At the Third Academic Convention of the Chinese Quarternary Society held in Beijing in 1979, eighteen papers were read on the subject. The Institute of Geography in Beijing has a pollen laboratory for Quarternary climate investigation by spore and pollen spectra interpolation. Deep drilling of lakes is being undertaken to obtain continuous records of climatic changes during the last two million years.

Physical climatic research in China was pioneered by the late Zhao Jinzhuang and others. Longstanding interest has been evinced in the Qinghai-Xizang (Chinghai-Tibet) Plateau. As early as in 1956 a state prize was awarded to a group in the Meteorological Laboratory of the Geophysics Institute, for work on 'the influence of the Tibetan Plateau on the general air circulation of East Asia and the weather in China'. (The Laboratory later expanded into an independent institute: see later.) The importance of the 'roof of the world' derives not only from its vastness – nearly a quarter of the national land territory – but also from its crucial effects, dynamically and thermally, on the summer monsoon circulation of South East Asia. The latest in a series of research programmes consists of field observations made during May–September 1979 followed by data analysis and laboratory experiments. Some of the results so obtained have been presented at a symposium in July 1982, at Kumming in Yunnan. The field work was conducted through 223 surface stations and 80 aerological stations, which provided 4 000 radar echo pictures as well as other data. The focus is on the outward movements of rain-bearing synoptic and mesoscale systems, many of which originate in the high planetary boundary layer over the Plateau, towards the eastern, southern or even northern parts of the country. These pressure systems were frequent occurrences and, if steered out of the Plateau, often result in severe storms or devastating floods downstream. It is, therefore, urgent to improve the understanding of their formation, evolution and movements, as well as the relationship between these phenomena and precipitation. Laboratory projects, concerning large-scale horizontal and vertical circulation patterns over the Plateau, are pursued at the Institute of Atmospheric Physics in Beijing. The work includes both numerical modelling and empirical simulation. The Institute houses a large rotating table on which observers can stand. A series of experiments have been performed since the mid-1970s in which the Plateau was simulated by a half-ellipsoidal body placed on the bottom of an annulus, and which may be heated, cooled, or neither heated nor cooled. Current experiments are providing insight into formation mechanisms of the circulation over the Plateau and its surroundings, for example a huge anticyclone in the upper troposphere during the summer.

Another research centre that has contributed significantly to the understanding of the climatology and meteorology of the Plateau is Lanzhou Institute of Plateau Atmospheric Physics. Both are subordinate to the Academy of Sciences and will be given unified descriptions at the end of this section.

Short-term weather forecasting is nationally coordinated in Beijing by the Central Meteorological Bureau, an agency directly responsible to the State Council. A recent development in the activity is synoptic dynamic modelling, a procedure that combines synoptic experience with the theory of dynamic meteorology. The Bureau also prepares 10-day and monthly outlooks of mean conditions, using trend charts, circulation indices as well as methods of linear regression and spectral analysis, apart from climatological records. Numerical forecasting has evolved into utilizing *ab initio* equations of increasing sophistication, and emphasis is now also placed on the statistical-dynamic approach, in which physical quantities appearing in the dynamical equations are treated statistically wherever they are difficult to determine exactly. Verification statistics of all these forecasts have consistently been good by world standards. Prediction success probability is around 0.7 overall, being higher for normal conditions but lower when abrupt weather changes are due, except in the case of typhoons, the paths and evolutions of which can now be fairly accurately foreseen.

Weather centres at the provincial, district and local levels are supervised by governments at the corresponding level, although communicatively they are all closely linked to the Central Bureau. The organization of meteorological services at the local level is a good illustration of how 'science for the people' is actually implemented. Not only local provision exists in the form of permanent weather posts, but also provincial centres send down technical personnel to communes or fishing fleets in times of special need. The practice of 'science by the masses' is realized in that many local posts are largely, if not entirely, run by peasants or fishermen who have either been trained or have educated themselves in the basics of meteorology. Some 'fringe' methods are in daily use at these posts, exemplified by the observation of clouds or animal behaviours; also the temperature of the soil a few metres deep may be monitored and taken into account when prediction is made on the mean precipitation for the following season. China has one of the finest synoptic networks in the world.

Apart from routine forecasting activities, research is being fostered on weather modification, cloud physics, radar meteorology and remote-sensing instrumentation.

Hail suppression and rainfall enhancement are ongoing projects which were started in the 1950s in many districts. A small group at the Institute of Geography is investigating hailstorm formation. Lanzhou Institute of Plateau Atmospheric Physics directs a large programme on hail clouds. The Institute of Climate run by the Central Meteorological Bureau has discontinued work on rainfall enhancement by dry ice seeding, but since the mid-1970s has sponsored trial seedings of warm cumulus clouds with pulverized salt released near the cloud base. Cloud physics occupies a prominent place at the Institute of Atmospheric Physics which, as noted below, has a Laboratory of Cloud and Precipitation Physics as one of its seven divisions. An example of the work in progress there is the study of the role of electrostatic interaction in the observed rapid growth of water droplets from 10 to 50 microns radius, a process which theorists in the West cannot fully account for by other mechanisms. At the Geophysics Department of Beijing University, experiments have been carried out on the formation of ice droplets, in particular at low degrees of supercooling when freezing rates observed are much faster than expected from standard theory. China is known to be developing laser radars or 'lidars' as cloud base detectors. The instruments may later be utilized in environmental protection, to detect atmospheric aerosols through the Mie scattering effect. Tests are also being made on acoustic radars, which can provide information about the vertical distribution of temperature by probing localized temperature gradients. As far as the conventional type is concerned, a national network of meteorological radars has been established. Excepting a few 5-centimetre radars of Japanese make, they are of indigenous manufacture and of two types: 3-centimetre and 10-centimetre operating at pulse rates of 400 and 200 Hertz, peak powers of 75 and 1 000 kilowatts, and ranges 300 and 700 kilometres respectively.

Large-scale pictures of cloud cover are regularly obtained from Japanese and American satellites, at weather centres and research departments in Beijing and other cities, by the use of receiving sets serially produced in the country. The Institute of Atmospheric Physics has, since 1973, received broadcast signals from the Very High Resolution Radiometers on the NOAA series of satellites. The 1697.5 Megahertz signal, collected by an automatically steered 2.5-metre dish, passes to an *in situ* constant (room)-temperature parametric amplifier incorporating a WB-51 GaAs varactor diode. It is then down-converted to 10.7 Megahertz and pre-amplified, before being sent indoors to an IF amplifier, an amplitude limiter, and a demodulator of the dynamic tracking filter type. Finally the FM output is fed to a display unit of the roller-scan type. Images produced in both visible and infrared channels attain a resolution of 0.85 kilometres at the satellite subpoint. The Institute, as well as that of Remote Sensing Application, have been experimenting with infrared remote sounding, microwave radiometric choppers and telemetry devices. Thus, the fact that no meteorological satellite is planned to be launched soon is more a matter of political priority than a reflection of inadequate technical competence.

The Institute of Atmospheric Physics, formerly the Laboratory of Meteorology in the Institute of Geophysics, was formally established in 1966. It comprises seven laboratories, an observation station (the Beijing Meteorological Tower) and a computer centre. The laboratories carry respectively the titles: General Circulation and Long-range Weather Processes, Dynamic Climatology, Atmospheric Dynamics and Numerical Weather Prediction, Atmospheric Turbulence and Diffusion, Severe Convective Storms, Cloud and Precipitation Physics, and Atmospheric Remote Sensing. Some of the more striking findings by their groups relate to South East Asian monsoon circulation, adjustment processes of rotating atmospheric motion, the interaction of sea and atmosphere and its relevance to long-range weather forecasting, as well as atmospheric pollution in mountainous regions and associated meteorological research. Significant advances have been achieved in the dynamics of cumulus cloud, theory of precipitation of warm clouds and study of rainstorm formation processes in China. Progress has been made in atmospheric sensing by means of lasers, in the development of telemetry systems for the Beijing Meteorology Tower, and in instrumentation and data analysis at meteorological satellite receiving stations.

The Lanzhou Institute of Plateau Atmospheric Physics, founded in 1974, contains the three divisions of Plateau Meteorology, Cloud Physics, and Atmospheric Sounding, plus a group on Boundary Atmospheric Physics. Important contributions from the first division include the evaluation of the thermal and dynamic effects of the Qinghai-Xizang Plateau topography on atmospheric motion, greater understanding of dynamic and thermal boundary layers, and improvements upon numerical forecast models to take topographic effects into account. Personnel from the second division have determined discrimination methods for hail clouds and characteristics of cumulonimbus lightning and electricity. Recent activities in the third division cover the design of lightning flash counters, field counters, field mills and range-calibrated equipment for radar.

All in all, the Chinese are strong in climatology and meteorology although gaps remain in a few places. One weak area is numerical prediction. However, this relative backwardness is due more to a deficiency in computer power than in scientific expertise. A great improvement is expected when there is a provision of better mainframes, peripherals – especially graphics terminals – and, last but not least, software (Chapter 8.5).

5.2 Oceanographic Disciplines

According to the 1978–85 Science Plan oceanography is among the twenty-seven spheres of research to be encouraged, although it is not included as one of the eight top priority fields. Activities within its scope have been conspicuous in recent years. A comprehensive survey of the sea between China's Zongsha, Xisha Islands and her Nansha Islands was completed by the ship *Experiment* in 1977. The next year witnessed the survey of the East (China) Sea. The most recent excursion was in 1982 when a State Oceanographic Bureau research vessel of 10 000-tonne deadweight brought to a close its first long-range cruise in the Pacific. Chinese eminence in the sciences as well as the practical affairs pertaining to the ocean may soon be a reality. Incidentally, from May to June 1983 the Chinese navy carried out a training cruise, the furthest so far, in the western Pacific. The discussion below will follow the delineation into physical, geological and biological oceanography. For completeness fresh water fishery is also included.

In physical oceanography, one of the topics enjoying much attention is storm surge and tidal prediction, a subject clearly of practical importance. Storm surges are investigated at the Hangzhou Institute of the State Oceanographic Bureau, through the application of regression methods to weather data collected by the Central Meteorological Bureau and to tidal records from tidal gauges. The spectacular Qiantang bore on the Hangzhou Estuary has been of special interest. It reaches a height of 8 metres and a speed of 15 knots; its power potential has been estimated at 5 gigawatts, although no plan seems to exist for exploiting it as a tidal energy source. The Shandong College of Oceanography at Qingdao is a centre of expertise in storm surge studies using regression and dynamic methods, as well as on the prediction of tides in shallow water by both geographical and numerical modelling. The Institute of Oceanography, also at Qingdao, similarily encourages research in tidal prediction – one of the methods employed being quasi-harmonic analysis – and in the attempt to explain frictional effects on tidal bores and tidal dissipation in the Yellow Sea. Another group is examining waves and wave forcing. The work includes basic study of the dynamics of surface and internal waves, apart from instrumentation development, efforts which are all related to offshore oil drilling operations (Chapter 7.2). A third group is engaged in hydrography of adjacent areas. In particular, it is looking at summer currents of cold water in the Yellow Sea, estuary mixing of the Yangtze River and the Kuroshio system. The last named circulation system, being the Pacific counterpart of the Gulf Stream, flows past the east of Taiwan Island and the south of Japan to join the North Pacific Current. The Institute is under the administration of the Academy of Sciences and will, hereafter in this section, be referred to as 'the Institute of Oceanography'. Towards the end of the section an integrated account of the scope and functions of its Physical Oceanography groups as well as groups in its other divisions will appear. The Kuroshio is investigated also at the Hangzhou Institute. Recently, the Institute of Mechanics (see p 181) installed a one-metre diameter rotating table, with the aid of which the Kuroshio system is being simulated in a topographic model of the near-coastal ocean. Another apparatus for the study of geophysical fluid dynamics, a water tunnel for the determination of velocity distribution by laser anemometer, has taken shape and experiments are being initiated.

Shallow water circulation is investigated by the 'Dynamic Factors in Ocean' group at Dalian Institute of Technology, relying chiefly on mathematical modelling.

Oceanic geology, thanks chiefly to its importance to offshore petroleum exploitation, has gained much ground in the last decade. Seismological surveys of potential oil beds are mainly, but not exclusively, the responsibility of the Marine Geophysical Prospecting Division of the Ministry of Geology. Chinese equipment is still inferior to the best in the West; in the South Sea survey (see p 124) they yielded information only down to 4 000 metres of sediments. Shallow seismic reflection profiling is carried out at the Institute of Oceanography to monitor sediment thickness in harbours. The South Sea Institute of Oceanography at Guangzhou, likewise an Academy of Sciences institution, devotes some effort to submarine geology, tectonics and geophysics. The Institute of Oceanography is also engaged in mapping magnetic anomalies in the East (China) Sea. Charts for the South Sea have been produced by the Guangzhou branch of the State Aquatic Products Bureau. The most widely used magnetometer is the proton precession type made by the Geological Instrument Factory in Beijing (see p 125). The best model has a sensitivity of 0.1 gamma and its specifications compare favourably with those of the world's leading makes.

Research in marine geochemistry is scattered among various localities and will be considered in Chapter 5.4.

Biological oceanography is fairly well represented. Both the South Sea Institute and the Institute of Oceanography of the Academy of Sciences are active in coral research. The latter has done much work on algae: it is responsible for the success in selective breeding and radiation-induced mutation of the Laminaria kelp, parthenogenetic clones for the female gametophyte of which have been produced at Shandong College of Oceanography. Phytoplankton, zooplankton and benthic communities are investigated at all the three establishments, as well as at Xiamen University, the institutes under the State Oceanographical Bureau, and Yellow Sea Fishery Institute of the State Aquatic Products Bureau – in their respective oceanic regions. The last-named marine fishery institute, located in Shanghai, is as its name implies engaged in faunistic research and mariculture development. It has in the last few years accomplished the artificial breeding of a number of rare piscine species. Incidentally, a significant development is that Chinese marine biologists took part, for the first time, in the 1981 international programme for BIOMASS (Biological Investigation of Marine Antarctic System and Stocks). Although China has yet set up no permanent stations in either polar region, the National Committee for Antarctic Research was established in the same year. Since then scores of scientists on various missions dealing with oceanography as well as meteorology and geology have visited the continent. In 1983 the government decided to seek membership of the so-called Antarctic Treaty. Chinese geologists have also taken part in the 1982 Cambridge Spitsbergen Expedition, as will be mentioned below in section 5.

In the subject of fauna systematics, various research centres have become noted for work on specific families. Examples are the fishery institute just mentioned, for gobioids, tetraodontoids, scorpaeniforms, sharks and rays; the Oceanography Department of Xiamen University, for copepods; the Institute of Zoology subordinate to the Academy of Sciences, for clupeiforms, eels, myctophoids, mugillids, catfish and flatfish; and the Institute of Oceanography, for perciforms, penaeids and stomatopods. The Institute of Hydrobiology in Wuhan, again of the Academy, is an inland water fishery research centre renowned for the study of cyprinids. There are about twenty species of freshwater fish of economic importance in China, virtually all of which are members of the cyprinidae.

Broad descriptions of each of the three freshwater and marine institutes in the Academy of Sciences are now given here. The Institute of Hydrobiology, the first to be formally set up in 1950, is a descendant of the former Institute of Zoology which existed from 1934 to 1949. A multidisciplinary bio-scientific research centre, it has six divisions, namely Ichthyology, Fish Genetics and Breeding, Fish Pathology, Ecology, Phycology, and Water Pollution Biology, the last division having already been discussed in Section 5 of the previous chapter. It runs in addition a freshwater fish museum and an experimental fish farm. Over the years much work has been done on the morphology, taxonomy and geographical distribution of inland water species, cetacean (Chinese river dolphin, Lipotes Vexillifer Miller) and the phylogenies of cyprinid fish and freshwater algae. It is there where over 10 000 specimens covering nearly all of the 800-plus known species in the country's inland waters have been collected together. In connection with major water conservancy projects, extensive investigations have been directed to aquatic organisms of various rivers, lakes and reservoirs. Comprehensive technical measures were worked out that led to a laudable increase of the fish yield in the East Lake of Wuhan.

The Fish Genetics and Breeding Division has succeeded in propagating or hybridizing a number of aquaculture species through genetic, cytological or acclimatization techniques. The blunt snout bream, Megalobrama ambly-cephala – which under the name of Wuchang Fish was referred to in one of Chairman Mao's poems – was cultured to replace the grass carp, Ctenopharyngodon idella. Although fast growing, the carp breeds only in warm waters and swift-flowing rivers during the monsoon season and is difficult to rear when young. The bream now serves as the essential first-stage herbivore in polyculture systems. Pargenthetically, it is interesting to note that undesirable characteristics may sometimes be turned into advan-tages: a recent report from the Weed Research Organization in Oxford advocates introducing the carp from China into British waterways for the purpose of keeping them free from overgrown aquatic vegetation. The carp is a voracious plant eater but does not breed indiscriminately, so the danger of runaway propagation of the predator itself is minimal. Besides the herbivore, another necessary element in pond polyculture is a detritus or humus feeder. For a long time no such species was found suitable, until the hydrobiologists discovered and artificially bred the Chinese nase, Plagiog-nathops microlepis. The male mirror carp, Cyprinus carpio (an inbred strain of the common carp), has been crossbred with the female red carp, Cyprinus carpio haematopterus, and the male red carp with the female crucian carp,

Carassius auratus gibelio. The hybrid common carp and hybrid crucian carp are now widely farmed in many provinces. The grass carp and blunt snout bream, species which are not cladistically close, have also been crossed. In the Fish Pathology Division, the virus responsible for haemorrhage of grass carp was isolated recently. To this day more than ten farm fish diseases caused by parasites have been brought under control due to the Division's findings.

The Institute pioneered the technique of counting primary phytoplankton with carbon-14 radio-assay. It is also outstanding in research on the phylogeny of freshwater algae. Biology of blue-green algae have been extensively studied and several species with high nitrogen-fixing capacities screened for use as biofertilizers. The scope of environment-related work going on at the Institute has already been outlined on pp 89–90.

The Institute originally was situated in Shanghai and had a Department of Marine Biology separately located at Qingdao. When the move to Wuhan took place in 1954, the Department was expanded to become the Marine Biological Institute and then, in 1959, it was renamed the Institute of Oceanography. Its present nine divisions are: Physical Oceanography, Marine Geology and Geophysics, Marine Chemistry, Marine Botany, Marine Invertebrate Zoology, Marine Vertebrate Zoology, Experimental Zoology, Marine Instrumentation, and Information and Data. It operates three research vessels of 300, 400 and 1 000 tonnes deadweight respectively with two more about to be launched. Significant results have been obtained regarding the prevalent current systems, tides and tidal currents in the Yellow and the East Seas, the variations in structure of water masses, and the Kuroshio circulation system. Knowledge has been accumulated pertaining to tectonic evolution and historical sealevel fluctuations in the two oceanic regions, and formation of the continental shelf since the Quarternary. Sediment transport processes in nearshore areas and sediment composition in the East Sea are intensively studied; models are being examined that explain sedimentation of the East Sea continental shelf.

In biological oceanography, one excellent piece of work has been on the morphology, developmental biology and ecological characteristics of the seaweed Laminaria. Consequently cultivation and culture techniques were implemented that enhanced its iodine content, raised its production per unit area, and increased its temperature tolerance permitting farming to be extended south. Its genetics and diseases were investigated, the uses of the kelp diversified. Other recent achievements have been connected with the artificial rearing of fries and sprats, the cultivation of prawns, mussels and mullets, and the prevention of fouling by marine organisms.

The South Sea Institute of Oceanography, independently set up at Zhanjiang in 1959 but later moved to Guangzhou, comprises eight laboratories, an Information Department and three experimental stations. The stations are situated at Zhanjiang, from where the headquarter was moved in 1973, at Shantou and in Hainan Island. The research ship *Experiment* mentioned in the introduction to this section belongs to the Institute; two others are near completion. The eight laboratories specialize respectively in Physical Oceanography and Meteorology, Tectonics, Coastal and Estuarine Processes, Marine Sedimentation, Marine Physics, Marine Chemistry, Marine Biology, and New Techniques. They have made progress in areas of research related to the physical and chemical characteristics of sea water, survey of the mouth of the Pearl River, geomorphology and geology of coral islands, and ecology of island organisms in northern parts of the South Sea and in waters adjacent to the Xisha and the Zhongsha islands. Significant results have been obtained relating to systematic floristic study, pollution and environmental quality evaluation of estuaries and shoals, the physiology and ecology of pearl oysters and Porphyra, and the control of biofouling. A marine buoy system for automatic telemetry of sea meteorological data has been developed in collaboration with other institutes.

Finally, a brief sketch is given of the Shandong College of Oceanography, which is apparently the only higher educational establishment occupied exclusively with marine science. Besides taking in regular students it also admits people on short-term training and provides correspondence courses. Its departments bear the respective titles of Marine Hydrology and Meteorology, Marine Physics, Marine Geology, Marine Chemistry, and Marine Biology. It is a main supplier of standard sea water sealed in glass ampoules, which are needed for salinity, density and other calibrations.

5.3   Hydrology and Geomorphology

China has taken large strides in hydrology and geomorphology since 1949. The importance of these subjects to water conservancy and control has been recognized from early times.

Soil moisture is a central topic for hydrologists, since one third of a million square km or 28 per cent of the nation's arable land may be regarded as arid, especially is this so in the northwest. Thus, the Comprehensive Physical Geography Division at the Institute of Geography devotes particular effort towards the prediction of soil moisture, as well as the examination of soil and biotic conditions and their distribution patterns. At the Institute thermal gravimetry and differential thermoanalytical techniques are employed to quantify water contents of soil and clay. Soil moisture is also investigated at the various research organizations for soil science controlled by the Academy (as well as the Academy of Agricultural Sciences). Work on glacial melt and permeation is quite advanced, that in Xizang being the most outstanding and having mainly been supported by the Institute of Glaciology and Cryopedology at Lanzhou. Repeated terrestrial photography is used to observe glaciers, now supplemented by Landsat imagery. Much experience has been gained in dam site selection for hydroelectric projects with the help of aerial surveying (p 158).

As regards fluvial hydrology, since runoff coefficients, evaporation and discharge rates, and other important features of all major rivers have been measured, emphasis has shifted to research on the dynamic behaviour of groundwater. The Institute of Geography accommodates groups devoted to land analysis for the purposes of maximizing water use and maintaining navigable waterways, to sediment accumulation in reservoirs, and to aggradation of river channels headward from reservoirs. Its Hydrology Group uses both mathematical modelling and experimental simulation to study storm runoff and discharge in a small basin. The Changchun Institute of Geography (a general description of which, and of the aforesaid institutions, will appear below) contains a group which is responsible for mapping out the main waterways. The Soil Erosion Division, belonging to the Northwest Institute of Soil and Water Conservation which has already come up in Section 3 of the previous chapter, studies runoff mechanisms and control measures. This division has determined the zoning of soil erosion in the upper and middle reaches of the Yellow and Yangtze Rivers, published a chart of erosion types in the whole country, and analysed the sediment origins of small gully watersheds. The Soil Geography Division in the same place has compiled an 1:1 000 000 atlas of soil on the Loess Plateau.

Loess erosional characteristics constitute a major concern of Chinese geomorphologists, who also join forces with hydrologists to investigate fluvial processes. A recent development is the application of Landsat multispectral imagery to this study. Geomorphic work at the Institute of Geography focuses on aggradation in river valleys. The discharge of sediment along the middle and lower Yangtze is measured and related to the loess of central China. Thus, the relative erodibilities of the banks and the bed are being evaluated. Laboratory work at the Institute includes modelling the Lower Yangtze with the aid of a 600-square metre flume, which is adjustable for gradient, flow volume and sediment load. This experimental programme is connected to a massive, long-standing project to divert Yangtze water northwards along the Grand Canal.

Sedimentation in the Yellow River is a focus of enquiry at several institutes, the end being to help solve the vital problem of irrigation and flood control. Work related to the control and the harnessing of the Yellow and the Yangtze Rivers has been designated as a priority area as early as the 1956 Twelve-Year Science Plan. The Yellow River basin, the cradle of Chinese civilization, is still vitally important because a large proportion of its three-quarter million square kilometres is, in modern times, loess terrain under intensive cultivation and heavily populated. In summer the sediment content carried by the river in the lower reaches may climb to 1 500 kilograms per cubic metre or 0.5 by volume fraction: understandably, flood control is an urgent matter. The problem is examined at, for example, the Water Conservancy and Hydroelectric Power Research Institute. This organization keeps two hydraulic laboratories in Beijing's suburbs. The northern one, provided with a 1 300-litre per second water supply, is mainly used for sedimentation experiments; the southern one, for structure and materials studies. A recent contribution from the former is the theoretical evaluation of vertical diffusion coefficients for sediment particles of various sizes in turbulent flows characterized by different hydrodynamic parameters: the results agree well with empirical data. Incidentally, another division (Water Resources) is conducting an analysis of the nationwide distribution of rain storms.

A big group in the Northwest Institute of Hydrotechnical Research at Wugong, belonging also to the Ministry of Water Resources and Power, concerns itself with the laws of resistance, pulsation behaviours and the cross-sectional profile of velocity distribution specific to hyper-concentration flows as distinct from clear water flows. It performs flume experiments and compares the outcomes with field measurements. Conventional wisdom held that the large quantity of silt carried by the Yellow River was a result of soil erosion in the Loess plateau covering an area of half a million square kilometres along the middle and upper reaches. However, recent work at the Department of Water Conservancy Engineering, Qinghua University, has shown that the silt is composed chiefly of coarse sand from only two districts along the middle reaches. The districts extend less than 50 000 square kilometres so that sediment control is less difficult than was feared. The Department is continuing its research, both theoretical and experimental, on sediment transport, to learn, in particular, why sediment concentration is approximately uniform over the cross-section of the Yellow River, except in a clear layer a few millimetres thick at the surface, and why the larger sand particles in the sediment do not fall to the bottom.

Geomorphological investigations of swamp wasteland, with the aim of developing methods for reclaiming swamps through well-drainage and well-irrigation, is a speciality at Changchun Institute of Geography. The Institute named after Nanjing has a research group working on lake sedimentation, geomorphology of canals and water-soil conservation in mountainous areas.

Quarternary studies, much in evidence, are mostly devoted to Pleistocene and Recent glaciations. In this respect, the region that has recently received the greatest attention is the Qinghai-Xizang Plateau: it has the distinction of being the most extensive area of mountain glaciers in the world. Many of its 1 000 lakes are found to have been formed by glaciations, which recurred four times during the Quaternary. Another discovery is that some of its present 18 000 glaciers show uncommon, 'maritime-type' characteristics and movements (surges). It is also the largest region of periglacial geomorphology in China. The Institute of Glaciology and Cryopedology, with the cooperation of other units, is looking at permafrost problems, especially those relevant to the railway construction now proceeding on the plateau (p 173). An investigation has just been completed into the occurrence and development of mud-and-stone flows, which are relevant to the maintenance of the Xizang highway. Also completed, at Chengdu Institute of Geography, is a study of mudflows on the northern section of Chengdu-Kunming Railway. Incidentally, at the pollen laboratory of the Institute of Geography – a laboratory which we have come across in Section 1 of this chapter – pollen analysis is undertaken to identify Quarternary strata serving as pointers for groundwater locations in the Yangtze basin.

Another matter of research should be raised. It is on the unique karst topography, which extends over half a million square kilometres in southwest China, and deals primarily with the difficulties of water supply and management under conditions of high rock porosity. The Institute of Geography, for example, specializes in the practical problem of leakage of reservoirs. The Ministry of Geology and Minerals administrates an institute at Guilin devoted wholly to karst geology. Desertology is also actively pursued but, having been discussed in Section 4 of the last chapter, will not be considered here. Deserts, found mostly in the northwest, cover almost one-seventh of the country's land area, so that the control of desertification as well as the use of water (and soil) resources in sandy or reclaimed regions are legitimate concerns of Chinese hydrologists.

The Institute of Geography consists of ten research departments. Besides those for comprehensive physical geography, hydrology and geomorphology, there are divisions for Climatology, Chemical Geography, Economic Geography, Palaeogeography, Cartography, World Geography and Technology for Geographical Research. Alongside conventional studies of hydrology, geomorphology and biogeography, some of which have been referred to in foregoing paragraphs, enquiries are being conducted into the migration and transformation of chemical elements on the Earth's surface. The emphasis is on clarifying the exchanges of water, chemicals and energy in the environment on a region to region basis, so as to determine potential agricultural productivity. The consideration now extends to toxic chemicals; this aspect has been elucidated in section 5 of the previous chapter. The second group of the Institute's activities centres on the investigation of the interrelationship between man and his environment, with the aim of providing a theoretical framework of human geography in China. Overall, the central theme is to construct a scientific foundation for a master plan for the utilization and conservation of the natural environment.

The same theme governs the work of the Commission for Integrated Surveys of Natural Resources which, similarily, has specific departments devoted to technical problems associated with water and land resources.

Xinjiang Institute of Geography deals with the geomorphology, water resources, ice-snow phenomena, agricultural and animal husbandry conditions, and economic geography of Xinjiang province. Particular attention is focused on the physical geography of arid land, rational regional planning of field production and precaution against snow hazards. There are five research laboratories and one alpine research station, namely: the Water Resources, the Snow and Ice, the Quarternary Geomorphology, the Economic Geography and the Remote Sensing Laboratory, together with the Tianshan Avalanche Station.

Both regional geography and the study of the swamps fall within the scope of work at the Changchun Institute of Geography. It has five divisions: Northeastern Chinese Geography, Swamp, Chemical Geography. Remote Sensing Application and Cartography; an experimental cartographic reproduction workshop is attached to the last named division. Contributions have been made towards research on agricultural regionalization of Jilin province, urban planning and industrial layouts in the northeast, and pollution and its prevention on Songhua River. A navigation atlas of major national waterways has been compiled. A better understanding has been achieved regarding peat resources in China's swamps, comprehensive aspects of swamp wasteland on the Sanjiang Plain, and practical methods of reclaiming swamps through well-drainage and well-irrigation. In chemical geography, vital data have been collected on the relationship between local diseases and regional variations of the environment.

The Nanjing Institute of Geography was founded on the old site of the Institute of Geography when the latter moved to the capital in 1958. The premises now comprise the Limnology, the Geography and the Cartography Departments. The first is engaged in basic research on lake sedimentation, air-water interaction, productivity of water bodies, lake resources exploitation, environmental protection, etc. Agricultural, urban and environmental geographies and historic climatology are among the current subjects studied within the second division. The Department of Cartography is concerned with the projection of special maps and the processing of remote-sensing imageries. Since its establishment this Nanjing Institute has carried to successful conclusion many large programmes, including agricultural regionalization of Jiangsu province, general survey of land utilization and water-soil conservation in the mountainous areas of southern Jiangsu, geomorphological investigation on canals, comprehensive study of Taihu Lake, and research on sedimentary facies of Cretaceous in the Songliao basin. Besides the last two of these programmes, the analysis of Chinese historical climatic records, assessment of environmental quality in Nanjiang and vicinities, and examination of geographic distribution and environmental factors of liver cancer are among the on-going projects.

The descendant of the Institute of Geography's Southwestern Branch, the Chengdu Institute is the newest provincial geographical institute to be set up (in 1978). In the past in common with other regional centres of research, the domains of its work have been physical geography, geomorphology, economic geography and cartography. Special attention is now given to studies of mudflows and the physical environment of mountains. The work on mudflows focuses on the local conditions of their formation, their regularities of motion, mechanical properties and prevention or control. The topographical study of mountains stresses the laws of regional differentiation linked to vertical zones in mountainous areas, post-Neogene evolution and active features of the geographical environs. Six divisions are under the Institute. The first is oriented towards research on mudflows; the second, on landslides; the third, on mountains and agricultural geography; and the fourth on the application of remote sensing. A fifth division manages books and reference materials, whereas the sixth is responsible for laboratory experiments and data analysis. Tasks accomplished in the last few years include the observation and control of mudflows and landslides in Heisha, Dongchuan and Daying Rivers, along Chengdu-Kuming Railway, on the Chinghai-Xizang Plateau and in strongly seismic areas. Advances have been made in the work on the subdivision of agricultural regions, the geographical distribution of Keshan disease and the effects of zinc on soil in Sichuan province. Progress is being made in the assessment and reduction of sulphur pollution by natural gas in Weiyuan, and the use of remote-sensing techniques to monitor landslides.

The Lanzhou Institute of Glaciology and Cryopedology, an offspring of the former Lanzhou Institute of Glaciology, Cryopedology and Desert Research (section 4, previous chapter), became a separate entity in 1978. It has three research divisions – on glaciology, permafrost and mud-rock flow – as well as three technical divisions: in surveying and mapping, remote sensing and telemetry, and analysis of material composition. Its work concerns the existing glaciers in Qilian Mountain, Urumqi River of Tianshan Ranges, Batura of Karakoram and the Himalayas, and the ancient glaciers, avalanches, snow drift and atmospheric radiation balance on the Qinghai-Xizang Plateau. Also, a wealth of data has been acquired on the basic characteristics and physicomechanical properties of permafrost along Qinghai-Xizang Highway and in Qilian and Da Hinggan Mountains, data which are essential to the maintenance of the Highway and to the design of mines and communication engineering projects. In 1981 a group from the Institute collaborated with a German team in a field study of glaciers on Mount Jishi which, known also as Mount Anymaqen, stands as the gateway to the Qinghai-Xizang Plateau.

5.4 Geochemistry and Geomagnetism

Chinese research associated with continental oil and gas deposits will be considered in the first section of the following chapter.

Research in marine geochemistry is scattered among a number of places. Systematic work on water column sampling for oxygen, nutrients and hydrocarbons together with projects in mineralogy and geochemistry of sediments are undertaken at institutes under the State Oceanographical Bureau, Shandong College of Oceanography, Oceanography Department of Xiamen University, the Institute of Oceanography and South Sea Institute of Oceanography. Marine organic chemistry is the concern of one of the nine divisions constituting Guangzhou Institute of Chemistry. It was mentioned, in Section 6 of the previous chapter, that a group at Shandong College of Oceanography has been conducting hydrocarbon pollution surveys. Another group there studies oil reservoirs by taking drill core samples, using analytical techniques like X-ray diffraction, X-ray emission spectroscopy, thermal gravimetry and differential scanning calorimetry. The technique of induced thermoluminescence is employed additionally at the Institute of Oceanography. Serial production of X-ray sources and diffraction cameras has long been going on in China, but many spectrometers and thermoanalytical instruments are still imported.

Salt lakes have gained much attention: the government is keen to exploit their chemical resources. The Qinghai Institute of Saline Lakes has been set up for the express purpose of gathering basic knowledge for the comprehensive development and utilization of such resources. Founded by the Academy of Sciences, it merged in 1966 with the Institute of the Comprehensive Utilization of Saline Lake Resources from the Ministry of Chemical Industry, but remains under Academy administration. It is divided into seven departments, engaged mainly in research on physical chemistry of concentrated salt solutions, geochemistry of saline lakes, comprehensive utilization of their resources, mining methods for their deposits, analytical chemistry of salt-forming elements contained in them, and related inorganic and isotope chemistry. Since its establishment the Institute has enriched the knowledge about the distribution, origin, salt-forming evolution and microgenetic laws of saline lakes in the Qinghai-Xizang Plateau region. Its staff have made many refinements to mining methods of solid and liquid deposits from these lakes, and to the technology for separating and extracting lithium, potassium, boron, bromine, iodine, rubidium and caesium salts. For example, a way was worked out in which sulphuric or hydrochloric acid can be replaced by chlorine water in the manufacture of iodine. Many novel techniques have also been evolved for the analyses of salts and brines.

Little geochemical study appears yet to have been carried out on freshwater resources. However, large-scale hydrological developments on the Yangtze (p 157) and the Huai Rivers are in the pipeline. In parallel with the industrial developments of the basins of these rivers, a systematic geochemical survey of all main waterways is planned. The Department of Oceanography in Amoy University has initiated some research in this direction. Preliminary results indicate that the chemistry of the seven rivers studied is dominated by the weathering of carbonates and evaporites, with significant contributions from the degradation of aluminosilicates. China's freshwater area amounts in aggregate to nearly twenty million hectares. About a third of that area can be used for fish culture; of this fraction some 70 per cent is in actual use. Chinese research in aquaculture has already been highlighted in section 2. Work of a chemical nature is in progress at a number of institutes, as outlined previously in the chapter on Environmental Sciences.

One particular geochemical property of groundwater, namely the random concentration, is the subject of many field and laboratory projects, due to its importance as a parameter in earthquake forecasting. Boreholes are drilled to reach basement rock at the end point of a fault to serve as observation wells for the area in question. The groundwater is pumped up and let to pass through a trap, through which air is bubbled to act as the carrier of randon to a detection chamber. The radioactivity of the accumulated gas is measured with either a Geiger counter or a zinc sulphide scintillation apparatus. In the laboratory, experimental and theoretical work are underway aimed at the edification of the relation between rock deformation and changes in the randon emission rate. Much of the efforts in this direction are directed by the Seismo-Geochemistry Group under the Geochemical Division in the State Seismological Bureau's Institute of Geology. There are three other groups under the Division. The one entitled Rock and Mineral Research studies petrology on a region to region basis, with the emphasis on typhonic and Cenozoic volcanics and rocks from both plate boundaries and deep fault zones. The Carbon-14 Dating Group is examining Quarternary stratigraphy in seismic areas, to establish a chronology for paleo-earthquakes and neotectonic activities since epi-Pleistocene. The last group is engaged in isotope analysis, with the focus on Cenozoic dating. The method most commonly employed by the group is potassium-40 argon-40 dating. The UHV hydrogen preparation system used was bought from Vacuum Generators Ltd, but the mass spectrometer, a MM 602D, is of

Chinese made and of high quality. A general description of the whole Institute of Geology will appear in the next section.

General geochemical analysis service is available in virtually all geologic institutions of the country, but work of a systematic nature is mostly supported at the Institute of Geochemistry. One of the ongoing projects there is the chemical characterization of granites in the whole of southern China. The Institute was formerly part of the Institute of Geology in Beijing but in 1966, on becoming independent, was relocated at Guiyang in Guizhou Province. It consists of twelve divisions in addition to a reference-information department, a computer laboratory and a workshop. The fields covered are sedimentology and organic geochemistry, ore deposit geochemistry, elemental and regional geochemistry, basic mineralogy, mineralogical physics, crystallographic mineralogy, mineralogical chemistry, mineral synthesis, rock-mineral analytical methods, experimental geochemistry and earth-interior geochemistry, cosmochemistry, isotope and nuclear geochemistry, and Quarternary and environmental geochemistry. Some representative work in isotope geochemistry and cosmochemistry will be mentioned shortly hereafter, and that in the mineralogical subjects, in Section 1 of the following chapter. The Quarternary Group has succeeded in defining the Sinian geochronological scale together with some Quarternary boundaries and in generalizing many evolutionary regularities governing Loess accumulation.

The Institute is generously equipped, possessing photon-counting X-ray diffractometers and optical spectrometers, a scanning electron microscope with semi-quantitative energy-dispersive X-ray facility, solid-source double-focusing mass spectrometers of both Chinese and foreign models, as well as Chinese atomic absorption spectrographic and electrochemical analytical instruments. High-resolution gas chromatography, neutron activation and isotope dilution techniques are also in use; the only conspicuous deficiency is the absence (as of 1981) of mini- or microcomputers interfaced to equipments.

At the Isotope Geochemistry Laboratory of the Institute the potassium/argon diluent technique is, whenever feasible, preferred in the determination of the isotope ages of crustal rocks. Methods based on uranium-238/lead-206, uranium-235/lead-207, rubidium-87/strontium-87 and carbon-14 are used but seem to be less popular. (In the State Seismological Bureau's unit mentioned earlier, the favoured techniques are those of potassium/argon and carbon-14.) For meteorites, dating by lead-207/lead-206 or fission track has been applied in addition to that by potassium/argon. The Institute took part in the multidisciplinary investigation on the spectacular meteorite showers that landed in Jilin Province during March 1976 and August 1978. A model for the evolutionary history of these meteorites has been constructed.

Meteorite study is also being continued by the Mineralogy Division of the Institute of Geology in Beijing – activities of the Institute as a whole will be described at the end of the next section. There Mossbauer spectroscopy has been utilized to map the distributions of iron in olivine and pyroxene. (The same technique is applied to the analysis of uranium ore composition by the

Applied Nuclear Physics Group of the Institute of Modern Physics at Lanzhou.) The Isotope Geology Division has made rubidium/strontium and uranium/lead datings of sediments baked from basalts and ashes found at Zhoukodian, the 'Peking Man' site. Thermoluminescence experiments on carbonate rocks and Quarternary Loess sediments have been performed at the Laboratory for Chemical Analysis of Rocks and Minerals. Another recent piece of research from this Laboratory is worth special mention, namely the chemical analysis of the clay used to seal the coffin of a corpse two millenia old, excavated in Hunan. X-ray diffractograms, TG and DTA thermograms, and SEM EDX spectra showed the clay to be hydrophyllite and halloysite. The project was part of a comprehensive scientific programme to examine this prominent archaeological find, said to have involved twenty thousand people. Like the interministry surveys of the Qinghai-Xizang Plateau (pp 92–3, 110 and 116) and the Jilin meteorite investigation just mentioned, such multidisciplinary efforts are nice examples of cooperation among scientists subordinate to different administrative authorities.

Geochronological techniques for dating sediments over geological time-scales include, apart from isotope analyses, the interpretation of palaeomagnetic data. A recent piece of work carried out at the Institute of Palaeontology and Palaeoanthropology, Beijing, involved age determination by the analysis of geomagnetic reversal sequences.

Geomagnetism is studied mainly in connection with archaeology, geophysical prospecting and earthquake forecasting. Nevertheless, at the Institute of Geophysics in Beijing, attention has been diverted to the phenomenon of geomagnetic micropulsation. There are theoretical efforts directed at modelling disturbances in the magnetosphere, and plans to measure micropulsations by recording Earth currents. More advanced a stage has been reached by the work on the secular variation of archeomagnetism. At the Institute, the magnetization of brick, tile, pottery and fired clay structure samples are measured with astatic magnetometers. Tests are performed on the stability of thermoremanent magnetism, including viscous remanant magnetization (VRM) test and AF demagnetization at 50 Hertz up to 15 kiloamps per minute, before the directions and intensities of the fields are determined with the double-heating (up to about 900 K) method. Good results have been obtained from samples from the Beijing area for the last 3 000 years and from the vicinity of Loyang for the last 2 500 years. An overall description of different types of work done at the Institute will be given on pp 116–7.

Fairly comprehensive geomagnetic surveys have been conducted both on land and over territorial waters, by various ministries for the purpose of mineral exploration. Some more basic investigations have been undertaken at various organizations in the Academy of Sciences. For example, the South Sea Institute of Oceanography analysed the data collected by the *Experiment* over the waters around Xisha, Zongsha and Nansha Islands. Besides indicating favourable regions for hydrocarbon prospecting, the results confirm that the magnetic anomalies in the area are mainly determined by the nature of the basement rock, known from drillings (section 2).

Palaeomagnetism is a topic of interest to the Academy's Institute of

Geology. An illustrative piece of previous work is VRM measurement of Cainozoic basalt groups in the vicinity of Nanjing. A paper published in 1981 discusses palaeomagnetic polarity determination of rock samples from the two banks of the Yarlung Zangbo River, from which data a velocity of under 5.5 millimetres per year is deduced for the northerly drift of the Indian Plate. A recent piece of research, completed in 1982 in collaboration with the Institut für Geophysik at Zurich, led to a new magnetostratigraphical dating of the loess deposits in north China. A Late Pliocene age of around 2.4 million years was assigned to the oldest sample examined. Palaeomagnetism is also studied at the Geotectonics Division of the State Seismological Bureau's Institute of Geology. Included in the work there is the mapping of polar-wander paths. Pre-Cretaceous results have been found to disagree with those from Canada, but post-Cretaceous data are in substantial agreement with those from Europe. The information is valuable to palaeotectonic study, and comparisons with present geomagnetic measurements provide a basis for earthquake forecasting. China maintains an excellent network of geomagnetic observatories. Eight are operated by professional personnel, where the local fields are monitored continuously and correlations sought with seismic activities. Nearly a hundred mobile observatories serve regions of high risk. Many types of modern magnetometers are in serial production, and these will be described on p 125.

5.5   Neo- and Palaeotectonics

The most characteristically Chinese school of tectonics is 'geomechanics', founded by the late Li Siguang (J. S. Lee) under whom China's first Ministry of Geology was established in 1952. Parenthetically, this government office was subsequently downgraded to become the State Geology Bureau but regained ministerial status in 1977; it is now called the Ministry of Geology and Minerals. Geomechanics attempts to interpret the change in the rotational speed of the Earth as providing the dominating mechanism for the formation of large-scale geological constitutions. This approach has been supported by experiments with paraffin wax or clay to model faults and foldings under rotational and tilting forces. Emphasis is put on the mapping of the principal stress fields responsible for producing the linear fault and fold patterns that cover much of the country. Geomechanical stress analysis has been claimed to be superior to methods based purely on traditional concepts, when applied to seismotectonics (see next section) and the prospecting of hard-rock minerals (section 1 of next chapter); moreover, Li has personally been given the credit of making possible China's breakthrough in oil exploration among continental deposits (same section of next chapter). For these reasons geomechanics has occupied an enviable position in the Chinese geoscientific scene. The Institute of Geophysics carries out much geomechanical research in its geomechanics division; the Academy of Geological Sciences (*ibid*) has an independent Institute of Geomechanics. Work at the Institute of Geology are more based on the geosyncline-platform viewpoint. This classical theory has been enriched by the hypothesis of 'diwa' (*ibid*). Lastly, since its inception in the West, plate tectonics has become in China the youngest school to attract a large number of proponents.

In connection with the examination of tectonic processes the study of Qinghai-Xizang Plateau has led to significant results. The Plateau has an extremely interesting structure and well developed stratigraphy, being made up of a basement of pre-Cambrian metamorphosed rocks and an overlaying complete stratigraphic sequence from palaeozoic upwards. Furthermore, as the most spectacular consequence of continental collisions in the world, it manifests hyperactive magmatic and metamorphic phenomena, as illustrated by Quarternary volcanism and at present frequent earthquakes in its northwest. Thus, tectonic processes there have been immense in both spatial and temporal spans: of all places, it is one of the best for research on the mechanisms of crustal movements.

Much has been accomplished. Most contributions have come from the personnel of the following organizations: the Institutes of Geology and Geophysics; regional geological bureau in Xizang and the Institutes of Geology and Plateau Geology subordinate to the Ministry of Geology and Minerals; the Changchun, Wuhan and Chengdu Colleges of Geology under the joint jurisdiction of this Ministry and that of Education; and Geology Departments in Beijing and Nanjing Universities. Field work has been undertaken in a series of multidisciplinary expeditions onwards from shortly after 1949. These large-scale cooperative programmes are noteworthy not only in the scientific context, but also as proving that Chinese bureaucrats or technocrats are capable of ensuring smooth coordination between separate ministries.

It is now almost unanimously agreed that the Indian Plate is pushing into and under the Asian Plate, so that for the last 0.1 million year the Plateau has been rising at an estimated speed of 10 millimetres a year or four times the current uplift rate in the Alps. The example of a palaeomagnetic determination of the horizontal drift velocity, carried out by the Institute of Geology, has already been mentioned in the last section. The rapid uplift has led to tremendous physical – and biological – changes during this period, and even during the Pleistocene. The crustal thickening is slowing down, and the question remains controversial whether such a north–south horizontal compression was still appreciable during the Quarternary. However, the view that the tectonic system is stretched along its other axis, particularly in its southern part, has been affirmed by Chinese aerial and field observations. Further confirmation of this suggested dominance of east–west extension came from a Sino-French study in 1980. The joint expedition located a large number of north–south normal faults, which are very young if not still active, and which sharply cut the glacial and post-glacial morphology. Complex patterns of tortional tectonics have also been developed between the north–south and the east–west stress systems as the results of the interplay of the two.

The Chinese distinguish four stages in the tectonic development of the Plateau from its beginning. These are pre-Cambrian geosyncline, Palaeozoic platform, Mesozoic-Eocene differentiation and final uplift since Neocene.

During the Palaeozoic the southern part of the Plateau constituted a disjointed tectonic unit, but the question whether the separation was by an ocean awaits clarification. That extensive subduction took place in Mesozoic-Eocene time has been inferred from the discovery of ophiolite suits, melange and glaucophane-schist; that no complete oceanization occurred is suggested by the presence of epicontinental sediments. Based on the probable absence of a complete oceanic basin, some geodynamicists at the Wuhan College of Geology have cast doubt on the commonly accepted picture (see above) that the Plateau resulted from a subduction-collision between two plates. They propose that, instead, the uplift since Neocene, forming coronal upwarping in tectonics and highlands in geomorphology, has been a consequence of grand block movements caused by 'deep seated thermodynamic and gravitational processes'. Whatever the cause, the Plateau became integrated only just before the Neocene. Efforts are being spent to recognize which fractures at the margins and in the interior represent suture lines.

Tectonic evolution from the beginning of the Phanerozoic onwards for the country as a whole has been looked at in detail. Three palaeoplates have been delineated and named after, respectively, Siberia for that in the north; China, in the middle; and Pacific, southeast. The plates are separated by the Junggar-Solon and by the Taiwan Longitudinal Valley sutures; the Yarlung Zangbo-Indus suture divides the Qinghai-Xizang Plateau of the China Plate from the Indian Plate in the southwest.

The nucleus of the China Plate is the Tarim-Sino-Korean Massif, which appears to be a platform or shield composed of Archaean and Proterozoic metamorphosed basement. The Tian, Bei, Yin and Changbai mountain ranges all belong to the epicontinental condillera of the Massif. It is being examined whether an oceanic slab existed in the south before Palaeozoic times and subsequently subducted below the Massif. The Chinese are also trying to confirm the existence of an Early Palaeozoic folded belt extending from Lishui in Zhejiang Province to Haifeng in Guangdong, and of a Late Palaeozoic one more adjacent still to the southeastern Pacific Plate. Considering the Mesozoic metamorphism of the Dananao Group in Taiwan and the widespread acid-intermediate intrusions and extrusions in coastal districts on the mainland, they have postulated a Mesozoic subduction along the Pacific coast of Taiwan.

Interests are being initiated in geological structures in the rest of the world. For example, visiting Chinese geologists at Cambridge University joined in its Spitsbergen Expedition of 1982. They studied slip faults in the Billefjorden Fault Zone of the Arctic region, in an attempt to gain certain information about the palaeoplate of North Atlantic.

The Institute of Geology was originally founded at Nanjing in 1950, but then moved to Beijing four years later. From it, in turn, three new research centres have sprung: the Lanzhou Institute of Geology and the Guiyang Institute of Geochemistry, as well as the Institute of Geology under the State Seismological Bureau. In the beginning, its work focused on the prospecting of useful minerals such as iron, manganese and phosphorus, and a geological survey of unmapped areas that led to the confirmation of Caledonian geosyncline in Qilian Mountain. Later, attention was gradually shifted to the following activities: the establishment as disciplines on their own in China of fast-growing branches of geological science, like geochemistry and seismological geology; the compilation of the first geotectonic map of the country on the scale of 1 : 4 000 000; the advancement of the theories of nuclear geology and sedimentary ore genesis through imbibition of terrestrial weathering products; the formulation of the theory of fault block tectonics (section 1, next chapter); the introduction of the subject of rock mass engineering geomechanics; and the syntheses of minerals and rocks. At present, emphasis remains on research into fundamental theories of geological science. Examples are a multidisciplinary study of the composition and structure of the lithosphere and the principles governing its geological evolution, an analysis of the origin and distribution of various kinds of petrographic and tectonic belts, and an understanding of Chinese geological evolution. Ten divisions constitute the Institute. They are Stratigraphy, Sedimentology, Geotectonics, Engineering Geomechanics, Geothermics and Geomathematics, Petrology and Laboratory for Determination of Physical Properties of Minerals and Rocks, apart from Laboratory for Chemical Analysis of Minerals and Rocks, Mineralogy and Isotopic Geology; some contributions from the last three divisions have already been mentioned on pp 107–8. The Institute also runs a museum for mineral and rock specimens.

The Institute was once transferred to the jurisdiction of the State Seismological Bureau for a period of time. The Bureau, on being officially formed in 1970, took over its control from the Academy of Sciences, until early 1978 when it split into two. The present emphases of research and the ten divisions of the part that was returned to the Academy are described in the last paragraph. The part remaining under the Bureau, henceforth known as the Institute of Geology of the State Seismological Bureau, is housed in buildings on adjacent grounds.

This institution under the Seismological Bureau has done much in tectonics and other geological fields. Its achievements include the compilation of a geotectonic map of Eurasia on the scale of 1 : 8 000 000 and the construction of a three-dimensional model of the deep structure in the Beijing-Tangsha region. It is now organized into seven divisions, with a scientific staff of some 30 senior personnel, 170 junior researchers and 150 technical assistants. The first division specializes in geotectonics. There are current projects in geotectonics, neotectonics, seismic fault investigation, remote sensing of seismogeology, fabric analysis, zonation of areas of high earthquake risk, and paleomagnetism; some results already obtained in connection with the last topic have been noted in the previous section. The second division deals with technomechanics and contains the five groups of Tectonic Modelling, Crustal Deformation, Crustal Dynamics, Earth Tide and Seismomechanics. The third, the Deep-Seated Structure and Materials Division, has interests in magnetotelluric sounding, magnetotelluric sounding instrumentation, converted-wave sounding, aeromagnetics, geothermics, gravity and focal tectonics. The Tectonophysics Division is subdivided into the Groups of High-Temperature and High-Pressure Tests,

Novel Techniques, Mathematical Modelling of Seismotectonic Dynamics, Seismogenic Structure Simulation, and Computer. The Geochemistry Division has already been described in the last section. The sixth division is devoted to the geology of seismic disasters and induced earthquakes and has four groups, in Soil Composition and Structure, Soil Dynamics, Earthquake Damage Simulation, and Induced Seismicity. The last division is responsible for collecting published data, graphic materials and technical literature from inside and outside the country, maintaining a library and running a permanent geological exhibition.

5.6   Earthquake Forecasting and Seismic Engineering

It is clear that the focus of attention at the Institute just described is seismotectonics, which relates the distribution of earthquake epicentres and magnitudes to neotectonic features. The Chinese maintain an impressive network of seismic stations, comprising professional stations in 1 210 counties set up by regional branches of the State Seismological Bureau and, at locations of high earthquake incidence, over 5 000 amateur stations where simple equipment is installed. They are, on the other hand, capable of manufacturing a whole range of high-performance laser geodimeters, quartz bar extensometers, quartz pendulum tiltmeters, strong-motion instruments, as well as long- and short-period (one-second) seismographic systems. The only relative backwardness concerns computer facilities. A recent development is the analysis of Landsat data to reveal neotectonic and rejuvenated faults, which are thought to be indicative of surface stress distribution. An understanding of the correlation of this distribution with the seismicity pattern is crucial in attempting long- to middle-term forecasts. In 1980, a 1 : 4 000 000 satellite imagery map of the seismic structure of the country was published by the State Seismological Bureau (section 2 of the next chapter).

Besides tectonic and surface stress analysis and geological evidence of active faults, factors considered by the Chinese in making long- to middle-term forecasts include local variations in crustal deformation, ground tilt, land elevation and sealevel, elastic-wave velocities of rock, and earth current or vertical component of the geomagnetic field (p 108). An additional and quite unique approach is archaeo-seismological investigation, which can yield information on zones of tremor intensity and on the periodicity and epicentral migration of earthquakes. The country has the longest continuous documentation on seismicity in the world, kept systematically since the twelfth century BC. The earliest record dates from 4 000 years ago. To go further back, the Chinese are looking for prehistoric earthquake sites by field work, aerial survey and satellite remote sensing. Such sites left their vestiges in geological and geomorphological features: faults, distorted lake sedimentary basins, fissures, level slips, liquefaction of sand layers and ejection of water and sands. For example, the Shanxi Seismological Bureau is studying a site near Linfen, which is 0.2 to 0.3 million years old. This work has been acknowledged to be as imaginative and meticulous as any done elsewhere, including the Quarternary study on the San Andreas fault of California.

To 'catch' impending earthquakes to within a day, their epicentres to within 50 kilometres, and their magnitudes to within half a unit on the Richter scale, the Chinese rely on evidence from geodetic and geophysical measurements, and of groundwater, foreshocks, earth lights and anomalies in animal behaviour. A paper presented at the International Symposium on Continental Seismicity and Earthquake Prediction, held in Beijing in September 1982, listed ten indicators of imminent continental quakes. They are itemized below together with supplementary information abstracted from a popular book written for amateur seismologists:

(1) A growing incidence of land faulting.

(2) A gradual drop in the geomagnetic intensity throughout the one or two months preceding the quake, that turns into a speedy rise in the last few days.

(3) In some cases numerous minor tremors occur, from several days to several months beforehand. The 1975 Haicheng event, which was the first quake that Chinese seismologists successfully predicted, was preceded by 600 foreshocks in one year. In contrast, for the Xingtai type of event, which represents gradual rather than catastrophic changes, the indication is abnormal tranquillity. Only three minor tremors (of magnitudes 0.1, 0.7 and 0.8) were detected in the three months before the Tangshan quake: since eight years beforehand the seismicity of the area had been steadily decreasing.

(4) The level of groundwater lowers and often increases again. A Chinese practice is to monitor wells, purpose-dug to over 3000 metres deep at sensitive points in the fault zone if necessary, for groundwater level, temperature and chemical composition. From a few months to a few days before the quake the wells may become dry and then Artesian, change in colour and muddiness, or acquire water containing gases.

(5) Sharp variations in the content of radon (p 106) and carbon dioxide, and of ions of calcium, fluoride and chloride, usually point to an impending quake.

(6) Marked decreases or increases in the resistivity of the earth may occur a few days to a few weeks beforehand.

(7) A few days before, seismographic signals often display large-amplitude fluctuations.

(8) The premonitory phenomena of unusual animal behaviour includes that of birds, cows, pigs or goats becoming restless, chickens refusing to enter coops, geese flying and rats running 'mindlessly'. Top of this list comes the behaviour of snakes and venomous snakes in particular, which congregate in ditches or emerge from hibernation during winter to be found frozen on the road. These phenomena take place a day or so before the quake and reach a climax about two to three hours before. They are thought to happen due to noise, produced by the cracking of underground rock but audible only to the animals' sensitive hearing, by the release of hydrogen sulphide gas, and by the rise of underground temperature which brings out hibernating reptiles. We note that controlled experiments are being

conducted under simulated earthquake conditions to test this hypothesis.

(9) Sometimes, underground sound audible also to human beings can be heard. It increases in frequency until just before the quake, when it stops abruptly.

(10) During an hour or so preceding the event increasing levels of electromagnetic waves are generated but disappear suddenly the moment the quake strikes. The emissions often manifest themselves by jamming radio communications, especially short-wave transmissions. We may add that similar emissions were reported in a Western journal in the same month as the Beijing Symposium, by a joint team from Sugadaira Space Radiowave Observatory in Japan and the Institute of Geophysics in Moscow.

All in all, the country is at the forefront in most scientific aspects of earthquake forecasting, and ahead of the world in some. An equally important, social aspect is the willingness of her populace in general to get organized and respond to predictions of impending quakes, in dire contrast to the indifference often shown by people in the west. Few parts of China can be regarded as essentially non-seismic. Over 3 200 destructive earthquakes have occurred since 1177BC whereas, in this century, on the average five quakes of magnitude five or greater on the Richter scale have occurred each year, and in the period of October 1949 to June 1966 ninety-five took place that caused injury or death. By concentrating resources (amounting, in 1982, to ten thousand scientists supported by a budget of 60 million yuan) and by mobilizing the broad masses the government has, with but a few exceptions, done well in the assessment and mitigation of earthquake risk since 1966, the year the anti-earthquake movement got started after being triggered by the devastation of Xingtai. The notable exception is the Tangshan disaster of $M = 7.8$ on 25 July in the sad year of 1976. Long- and middle-term warnings of it had been issued in 1970 and January 1976 respectively, but the imminent alarm failed to get out in time. Among the reasons for this failure were the facts that it occurred where no strong shocks had been recorded before, and that most of the immediate precursors were apparent less than a day in advance. Moreover, within a few months of it a series of earthquakes took place nearby, such as the one at Haicheng ($M = 7.3$), Helingeer ($M = 6.3$) and Dacheng ($M = 4.8$). Their early precursors all blended together and could not easily be separated out.

In recent years attention has been drawn towards the relation between earthquakes and sharp variations in the temperature and pressure of oil, gas and water issuing from oil wells. The first national meeting devoted to this topic was convened at Qingdao in October 1982.

The government used to put more stress on the timely provision of warnings and subsequent implementation of precautionary measures, than on the construction of seismic-resistant structures. In recent years, however, much work has been done on the field measurement of dynamic models of buildings, as well as the laboratory model testing and the mathematical analysis of structures in conjunction with major construction projects. In cities situated on earthquake belts many houses have been examined for antiseismic capability and reinforced if necessary, and also more earthquake-proof rooms have been built to serve as shelters when needed. A new set of

'Earthquake-Resistant Design Standards for Industrial and Civil Buildings (T J 11–74)' came into effect in December 1974. A design code for the seismic safety of river dams, the 'SDJ 10–80', was promulgated by the Ministry of Water Conservancy and Power in 1980. Seismic engineering protection for large-scale hydroelectric installations has been cited as a key area of research in the 1978–85 Science Plan.

Several research centres under the Ministry have programmes on the earthquake behaviour of dams, that combine in situ recording of earthquake response with dynamic laboratory tests of dam models and numerical modelling using finite element methods. Among institutions of higher education, the Dalian College of Technology contains a Seismic Protection Laboratory that has made notable contributions. A model superposition method has been worked out by which, based on statistical treatments invoking the earthquake response spectrum principle, the maximum acceleration, base shear and hydrodynamic pressure of a dam during seismic events can be calculated. Qinghua University in Beijing has a Hydraulic Engineering Department where photoelastic, strain-gauge and other measurements are being performed on dam models: the work is funded by the Ministry. In the other department for Structural Engineering, experiments are in progress to evaluate the earthquake strength of reinforced concrete slab column connections that are found in building constructions. In this field of seismic resistance of buildings, however, the most important research organization is probably the Institute of Engineering Mechanics. An ambitious project there is devoted to the masonry structure, which is the most common type of construction in the country. True-size houses, 1/10 scale models on a shaking table, and wall components are tested to destruction. The Institute is under the dual guidance of the Academy of Sciences and the State Seismological Bureau; an overall description of it will be given at the end of this section.

The Academy's Institute of Geophysics, like the Institute of Geology discussed in the last section, was founded at Nanjing in 1950 but then relocated in Beijing. At first, the Institute devoted most of its energy to establishing a nationwide network of seismographic and magnetic stations, which was subsequently transferred to the administration of the State Seismological Bureau. However, it still operates a regional network around Beijing, comprising 19 unmanned stations for the fast determination of nearby events. Telemetry is handled by a DJS-6 computer (100 000 OPS, 48-bit and 32K core memory) interfaced with a 30-channel A/D converter. In recent years it has conducted the following investigations: a magnetic survey of the whole country; the design and construction of a magnetograph (a magnetic theodolite) and a recording system for seismic sounding; study of seismic wave theory and earthquake origin; an identification of the lithospheric structure of China; and research on the distribution of high-grade iron deposits on the basis of crustal structural analysis (cf first section of following chapter). The Institute took part in the multidisciplinary programmes for the Qinghai-Xizang Plateau, concentrating its attention on the deep structure of the Plateau. At present it lays strong emphasis on research into the constitution of the Earth's interior, and the characteristics of various geophysical fields including the magnetosphere. Its seven divisions are: Theoretical Geophysics, Geomagnetism, Geodynamics, Geophysical Prospecting, Magnetospheric Physics, Earth's Crust and Upper Mantle, and Instrumentation and New Techniques. There is a department of scientific information and a high pressure laboratory for soil mechanics is under construction.

At the Academy's Institute of Acoustics, a small group has recently formed, to work on seismoacoustics. Its goal is to establish acoustic methods of earthquake forecasting.

As was just said, the Seismological Bureau now administers the national network of seismographic stations. Standard equipments at a permanent observatory in this network include the DD-1 short-period seismograph, the Model 513 intermediate motion instrument, and the RDZ-1-12-66 strong motion apparatus. The DD-1 records three components with flat response from 0.1 to 1 hertz. The RDZ, manufactured at Factory No. 581 in Beijing, usually provides for 8 horizontal and 4 vertical transducers, and gives a normal range of 5–100 gals and a flat response over 0.5–35 hertz. The best long-period seismometers allocated to the network offer stable measurements up to 70 s horizontally and 60 s vertically, and a peak magnification of 50 000 with the use of DC amplifiers.

The Bureau runs its own Institute of Geophysics. Like the Institute of Atmospheric Physics and the Lanzhou and Kunming Institutes of Geophysics, it is an offshoot of the Institute of Geophysics that has just been described. As in the case of the geological institutes (previous section), its parent organization and it are adjacent to each other (in a district between down-town and the geological institutes). It contains divisions devoted to (1) regional seismicity of the Beijing area, (2) seismicity of the entire country, including archaeoseismology, (3) geomagnetic precursory phenomena, (4) physics of the earthquake source, (5) study of deep crustal structure (to about 20 kilometres depth) from observation of small quakes, (6) microseismology and (7) instrumentation: improvement and innovation. Some 350 technical personnel work at the Institute. It acts as the national coordinator of all regional seismological departments, which are under the jurisdiction of the respective provincial authorities. It has been prominent in the sphere of international cooperation. For instance, information is exchanged regularly with the Short-Period Seismography Network of South East Asia under the auspicies of United Nations Development Plan and UNESCO.

The Institute of Engineering Mechanics is at Harbin. Since its reorganization in 1963 it has concentrated its attention on soil mechanics, earthquake engineering and related shock and vibration problems. It has its own network of strong motion seismographic stations. A newly added direction of research is structural stress analysis in relation to thermonuclear reactors. It contains the laboratories of strong motion network, engineering seismology, earthquake resistant structures, shock and vibration, soil dynamics, rock mechanics, composite materials, and vibration instruments and equipment. Its contributions include the drafting of seismic codes, assessment of damages to civil constructions after earthquakes, experimental and theoretical research into seismic protection of various types of structures, studies of ground motion characteristics and site effects, characterizations of the rheological properties of loess and timber, as well as the development of silicate materials and special kinds of concrete, such as the quick hardening, early strength, and radiation shielding varieties.

6 Mineral Industries

Chinese technical capabilities in geological exploration and extraction of minerals are the subjects of attention in this chapter. Processing and refining are also to be considered in the case of mineral metals but, for mineral and rock fuels, discussions will be deferred to the following chapter; the fertilizer mineral sector has already been dealt with in Chapter 6.3. The Ministry of Metallurgical Industry is responsible for the mining and refinery of metals except uranium and gold, regarding which the Ministry of Nuclear Energy and the Bank of China exert control respectively. For coal and petroleum there are separate ministries.

Mining and metallurgy share a long history in China: documentary and archaeological evidence shows that bronze, brass, copper and silver, gold, tin and zinc were mined and casted in times before 1 200BC. The iron age commenced soon afterwards. The ancient Chinese could grasp the idea of geological changes and even speculated about mineralization processes. The 'Book of Mountains and Seas', many parts of which have been dated as seventh to fifth century BC, contains various passages stating that where one ore is found another is likely to be beneath, an example being pyrite/alum. Not only, since such early periods, were searches conducted by these empirical associations, but also, in mediaeval age, bio-geochemical prospecting was resorted to. The 'Illustrated Mirror of the Earth' of the sixth century instructed readers that, 'If the stalk of (a certain) plant is yellow, copper will be found below'. The 'Miscellany of the Yuyang Mountain' published in AD863 asserted: 'Where in the mountains there is ginger then copper and tin should be near.' Iron and steel technology dates to 800BC. The casting of iron is one of the best-known example of traditional technology. Recent recoveries of ancient artifacts have indicated that by 750BC the Chinese knew how to pour iron into multiple moulds stacking on top of one another. This stack casting technique is still practised today, at for example the modern foundry of Foshan in Guangdong for the production of gears and precision metal parts.

6.1 Mineral Geology

Since the beginning of this century it has been well known that Chinese antimony ore tops all others in reserves and quality. China is also acknowledged to be the king of tungsten, being foremost in quantities of ore, annually produced metal and net export. Her kingdom certainly extends to cover the 'rare earths' – chemical elements that are hardly describable as 'rare' there – as well as to titanium. In the output of tin she lags considerably behind Malaysia but, in terms of reserves (about 2 megatonnes), takes second if not first place; the deposits in the Gejiu region of Yunnan province are of particularly excellent grade. The reserves in molybdenum, niobium and mercury rank second in the international league and all are generally of excellent quality. Identified deposits of iron and bauxite, totalling 44 300 and 1 100 megatonnes respectively, are abundant but often of low geological yields. High-grade iron ore needed for beneficiation is at present bought in substantial quantities from outside, and such dependence may persist for some time despite recent discoveries of good-quality deposits, so long as inland transportation remains a bottleneck. Indeed, China is planning to invest in Australian mines with the aim of securing an ore supply for the Baoshan Steelworks, an imported plant not capable of feeding on ore of even grades and which will be served by a port to be purpose-built nearby. In petroleum, the country has spectacularly changed from an importer to, since 1973, a net exporter of increasing world importance. Offshore oilfields are just beginning to be developed and they have been gauged by certain optimists to have the potential to rival those of the entire Middle East region.

All the aforesaid endowments, except the antimony riches, were unknown before 1949. The government's confidence and planning in the mineral sector would have been impossible had close attention not been devoted to exploration and exploitation throughout the last three decades. In 1981 alone it was reported that some 140 mineral deposits were discovered. It was also said that regional evaluations of rocks and minerals are basically accomplished, with the exceptions of high-grade coal and iron ores, so that emphasis has shifted to soil, groundwater and construction materials resources. Efforts concerning the first two resources are required because of the ever present need to raise agricultural production, as discussed in the introduction to Chapter 4, whereas the equally chronic, if less acute in consequences, shortage of housing underlines the urgency regarding exploration of the last. Parenthetically, it is surprising that partial reliance on traditional building materials and popular, collective participation in the housing process have not been strongly encouraged, if but for the purpose of alleviating the shortage. Interestingly, though, two factories for making building panels from rice and wheat straw by means of heat and pressure are under construction near Beijing; however, the technology involved came from a foreign company, namely Stramit International of the UK. The exploration for building materials has been under the direction of an independent ministry, which in 1982 was absorbed into a new Ministry of Urban and Rural Construction and Environmental Protection. A recent example of the work at the Institute of Geology under this Ministry is a mineralogical study on chrysotile of serpentinized dolomite type from mines in northern China, with X-ray and electron diffractography, infrared spectroscopy, thermal gravimetry and differential thermal analysis.

Even remote regions like Xizang have been surveyed for resources to some extent: a geological team announced in late 1982 the discovery of reportedly the world's largest titanium deposit on the Plateau. Such continued successes in locating ores have been attributed in part to fresh insights from geodynamics (see section 5 of previous chapter). The analysis of the northwesterly thrust of the Pacific Plate against Eurasia, according to the theory of fault block tectonics, indicates the formation of many weak zones in specific regions of China conductive to mineralization. Along this line, a group at the Institute of Geophysics (described towards the end of the last chapter), for example, has commenced investigating the distribution of high-grade iron deposits. Similarly the application of geodynamics, by emphasizing the aspect of dynamic evolution over that of historical origin in the search for geological structures conducive to oil accumulation, is responsible to a large extent for successes in the search. Petroleum source rocks in China are typically of continental origin: this characteristic was originally thought impossible by Western experts, who had been familiar only with oil beds found in extensive marine sediments. The Chinese locate candidate oil provinces by looking for depressed regions of what is now called first-order tectonic system, where thick sedimentary piles may have built up. Such philosophy formed the basis of the government's decision in 1958 to transfer some prospecting teams from the northwestern to the northeastern part of the country. Oilfields are delineated using the criterion of good reservoir conditions: their distribution is controlled by tectonic elements of second, third or still lower order. Shear and rotational shear structures are predicted to be the most preferred sites for hydrocarbon accumulation.

There are two major categories of oil-bearing structures in China: compressional basins elongated in an east–west direction in the west of the country, and tensional or extensional basins elongated in a northeast–southwest direction in the east. Included in the former category are the northern basins on the Qinghai-Xizang Plateau, the Tarim basin of southern Xinjiang and the Zungarian basin of northern Xinjiang, which are all intermontane or platform basins formed and preserved as the result of the northward movement of the Indian Plate against the China Plate (see section 5 of previous chapter). The tensional basins in the latter category were formed essentially parallel to the western boundary of the Pacific Plate along the Pacific coast of Taiwan Island. Since the westward subduction of the Pacific Plate has been effective in stopping the eastward advance of the China Plate, the offshore basins in the Yellow and the East (China) Seas, where hydrocarbon prospecting is gathering momentum, are narrow and elongated. At Zhongyuan Field bordering Henan and Shandong provinces, where oil production is being expanded at an accelerated pace, the oil-bearing strata are moreover often deep or broken, or both. The Sungliao basin, in the northern part of which the nation's first major oilfield to be discovered since 1949 (the Daqing: see p 148) is situated, and the basins of the North China Plain, on one of which the newer Renqiu Field lies, are intermediate in width. The northeast–southwest Sichuan basin where natural gas is being extracted is broader on the west, where it is furthest from the Pacific suture.

Petroleum geology has always been included as one of the responsibilities of the Ministry of Geology and Minerals. Its Institute of Geology in Beijing (see below) and the regional Northeastern Institute of Geology in Changchun took part in the discovery of the Daqing Field. It now has two more research establishments specialized in petroleum geology in onshore and offshore areas respectively (see same below). The subject is also studied intensively at the Institute of Scientific Research for Petroleum Exploration run by the Ministry of Petroleum. The Institute has laboratories in stratigraphy and palaeontology, sedimentology and mineralogy, organic geochemistry, and physical properties of sediment, such as porosity and permeability. The Sichuan Institute of Natural Gas Research under the same Ministry is credited with the clarification of geology structures on Triassic formation in the gas-rich province of Sichuan.

In China the single most important organization for geological research oriented towards the exploration of all kinds of minerals is the Academy of Geological Sciences sponsored by the Ministry of Geology and Minerals. The Academy has been formed from an amalgamation of both centrally and some locally directed institutes. Some of its activities necessarily overlap with the work of research establishments under other, more specialized ministries, but in fields like the compilation and publication of comprehensive geological maps the Academy has no competition. For ease of reference all the 18 institutes under the Academy are listed here:

Institute of Geology, Beijing

Institute of Mineral Deposits, Beijing

Institutes of Geology and Mineral Resources at Tianjin, Xian, Shengyang, Chengdu, Yichang and Nanjing

Institute of Geomechanics, Beijing

Institute of Rock and Mineral Analyses, Beijing

Institute of Plateau Geology, Chengdu

'562' Comprehensive Geological Brigade, Yanjiao, Hebei province

Institute for Research on Comprehensive Utilization of Minerals, Beijing

Institute of Hydrogeology and Engineering Geology, Zhengding, Hebei province

Institute of Karst Geology, Guilin

Institute of Geological Information, Beijing

Institute of Petroleum Geology, Beijing

Institute of Marine Geology, Qingdao

The National Geological Library and Geological Museum, both in Beijing, are also administered by the Academy. The staff of the Academy numbers over 5 000, among which about 3 000 are technical personnel. The last two institutes listed above are under joint jurisdiction of the Academy and the Ministry's Petroleum and Marine Geological Bureau.

The six Institutes of Geology and Mineral Resources identified in the list above are likely to increase in number before long. Each of them usually comprises the following divisions and related laboratories: Regional and Structural Geology, Palaeontology and Stratigraphy, Mineralogy and Petrology, and Mineral Deposits. Their work is strongly oriented towards local needs. Taken as an example, the Institute at Xian has collaborated with

Gansu and Sichuan provincial bureaus of geology to produce eleven large-area stratum profiles from the Xiqinling region. An Early Devonian epoch marine stratum was discovered, for which detailed palaeontological and mining assessment data were collected. The stratum is 1.2 kilometres in maximum thickness and has a transitional relationship with the Late Silurian epoch stratum below, whereas its angle does not correspond fully with the Middle Devonian above.

In addition, some related research is sponsored under the Academy of Sciences system. The Sedimentary Rocks Division of Nanjing Institute of Geology and Palaeontology conducts investigations on petroleum-, coal- and iron-bearing and phosphate sediments. The Guiyang Institute of Geochemistry has studied minerals from major mining bases in Jinchuan, Panzhihua, Mount Xihua and other places, analysed the regularities of ore-forming processes and developed a multi-origin metallogenetic theory. A result recently published suggests that the iron ore deposits at Bayan Obo and Shilu show surprising similarities in certain aspects, a finding that underlines the primary significance of studies on sedimentary environment in ore genesis and exploration. The Lanzhou Institute of Geology is concerned with the origin and distribution of petroleum. It has as its predecessor the Northwest Geology Laboratory of the Institute of Geology in Beijing, and contains one research group (Palaeontology) as well as four research laboratories (Tectonic Geology, Sedimentology, Organic Geochemistry, and Elements and Isotopes Geochemistry). Often in collaboration with its parent institute in the capital and with Xinjiang Institute of Biology, Pedology and Psammology, it has made notable contributions to many areas, outstanding among which is the hydrodynamics governing the development of inhomogeneous oilfields. It is continuing the work on the formation, accumulative conditions and evolution as well as the migration of continental oil and gas in erosion basins.

At the Academy's Changsha Institute of Geotectonics the focus is on similar aspects in the case of mineral deposits generally. Established as the successor to the Central-South Laboratory of Geotectonics and Geochemistry, the Institute conducts studies on metallotectonics and seismotectonics, being composed of the two divisions of Lithospheric Evolution and Metallogeny. Current work is concentrated on the origin and evolution of ore-controlling structures such as 'diwa', the geotectonic setting of formation and prognosis of the Late Paleozoic siderite deposits of southern China, and the tectonogeochemical characteristics and metallogeny of Nanling and adjacent regions. The term 'diwa' refers to crustal depressions and, as a new dynamic element introduced into classical geotectonics, was first formulated by a Chinese in 1956; that geologist is the present director of the Institute, which is prominent as the chief proponent of the resulting geosyncline-platform-diwa theory. It is now recognized that crustal evolution in China passed through the platform stage in the middle Mesozoic into the diwa stage, when crustal mobility was regained, as evident from the formation of volcanoes and subsided sedimentary basins. This period is thought to be a crucial stage for metallogenesis. Minerals deposited in diwa regions are both of inheritance and newborn in nature, and thus often yield polygenic compound ores. Geodome zones in such regions are usually endowed with nonferrous metals, whereas diwa basins hold the promise of coal, oil and gas finds.

From the universities a piece of work on igneous rock is worth special mention. Very recently a group in the Geology Department of Nanjing University completed a study on granites in southern China and proposed a theory, as well as practical techniques thus based, of finding tungsten and other minerals in them. Delineating them into the transformation, the symtexis and the mantle derived types, and their formation times into eight main periods from 1 400 to 70 million years ago, the geologists have correlated these types and formation times with their mineral contents.

6.2   Geophysical Prospecting

Geological-mineral maps of China have been compiled by the Institute of Geology under the Ministry of Geology and Minerals from 1958 onwards; for this contribution the leading scientists involved were awarded a Class I Prize by the State Scientific and Technological Commission. Geological, aeromagnetic and gravity maps of scale 1 : 1 000 000 are now in print for the whole of the country except some sea areas and outlying islands. Additional coverage on scales from 1 : 25 000 to 1 : 100 000 is available for nearly all of the interior provinces. Radiometric survey at 1 : 50 000 to 1 : 200 000 has also been completed for many regions, but uranium prospecting will be further discussed in section 4 of the following chapter.

Seismic surveys have been even more extensively carried out. They are conducted by the Ministry of Geology and Minerals, Ministry of Petroleum Industry, and China National Oil and Gas Exploration and Development Corporation, sometimes with duplication of efforts. Combined activities, both onshore and offshore, have amounted to 10–15 per cent of world exploration activities consistently over the last few years in terms of million metres drilled. The offshore sector used to be the relatively weak area, but much progress has been achieved recently. Hydrocarbon exploration in the Gulf of Bohai was, and that in the East Sea is being conducted entirely by Chinese workforce, albeit not exclusively with Chinese advisers and equipment. This operation is under the auspices of China–Japan Petroleum Development Corporation, the foreign partner providing some advanced hardwares. In the Yellow and South Seas, after carrying out preparatory airborne geophysical mappings during 1974–7, the Ministry of Petroleum Industry invited interested parties from abroad to cooperate in prospecting. For example, preliminary exploration of a region of the South Sea was undertaken by a Columbia University research vessel in 1980, and the survey subsequently completed by the Chinese ship *Experiment*, mentioned in section 2 of the last chapter. All data were analysed by Chinese geoscientists whose recommendations went to the Ministry. Outside participation in test drilling was allocated by area towards the end of 1981 and in early 1982 the China National Offshore Oil Corporation was set up to coordinate the entire operation. Types of oil sources identified have included stone bind, limestone, biogenetic limestone and palaeophylum reef. In the Gulf of Bohai, the oil-producing structures are chiefly tilted fault blocks created by tensional collapse in the Tertiary, reminiscent of those in the North Sea of Europe. China has several exploration drilling ships. The first, a catamaran fabricated from the hulls of two old cargo vessels at Hudong Shipyard in Shanghai, was launched in 1974 and called *Explorer I*. However, there remains clearly a deficiency in the number of deepwater drill-ships – and in modern seabed devices and down-hole equipment. Chinese preparedness for offshore field production and the more happy situation in the onshore sector will be examined in Section 2 of the next chapter.

In the country serial production of well-logging and isotope well-testing apparatus, gas analysers, seismographs as well as magnetometers, gravimeters and radiometers has been in process for a long time. The biggest supplier is Xian Geophysical Instrument Factory; the first domestically manufactured numerical seismograph came from there in 1976. When used in aerial surveys, magnetometers are suspended from planes or mounted on the wing tips. Many kinds are now made, including proton nuclear precession, saturation core, and alkaline vapour, both caesium and helium, optically pumped. Nuclear magnetometers of the top models such as the CHHK-2 have sensitivities of better than 0.1 gamma in the field (0.05 gamma in the laboratory). Semiconductor types are also available, having been originally developed at the Institute of Geology belonging to the Ministry of Metallurgical Industry. Under trial and evaluation is the dc SQUID magnetometer. One such type is fabricated by photo-etching double microbridges on a sputtered niobium film and provides a laboratory sensitivity of around 0.005 gamma. A screw double-point contact device is also being assessed. Field gravimeters are mostly of the quartz type, with specifications around the 0.02 milligal level and thus distinctively inferior to the best in the West, though some prototypes have been produced that offer a tenfold improvement in the sensitivity, for use in solid earth tide observation. A superconducting gravimeter based on the Josephson effect in an niobium Dayem bridge arrangement is under development at the Institute of Physics, Academy of Sciences, for application to the study of gravity changes prior to earthquake events.

Unlike the situation in many other disciplines, attention has not been focused on front-end instrumentation to the exclusion of data processing facilities. For example, minicomputers are installed in many places and utilized to make topographical corrections for gravity data. In induced polarization surveying, microprocessor-controlled receivers are being developed to record data which can then be processed on mainframes or, in the future, on dedicated microcomputers to give spectral IP parameters from complex impedance plots.

The implementation of remote sensing techniques on scientific and applied satellites is one of the objectives for accelerated research advocated in the 1978–85 Science Plan. Although China has herself placed into orbit a total of fourteen payloads (Chapter 8.2) none of them has been an earth resource technology satellite. Nevertheless, the task of developing one has been assigned to the Space and Technology Centre, located in Beijing and belonging to the Academy of Sciences. Founded recently in 1979, the Centre is composed of the divisions of Remote Sensing Application and Space Technology, as well as those of Space Sciences, Space Systems and Ground Network. Some activities of the latter three divisions will be mentioned later in Chapter 8.2. (In this connection, it may be noted that Shanghai Institute of Technical Physics belonging also to the Academy is a major centre of research on infrared technology. Instruments developed there include infrared horizon censors as well as airborne and spaceborne multispectral scanners.) Furthermore, not only has an indigenous earth resource satellite been planned, but also, since the early 1970s, transmissions from Landsats have been received locally. From 1973 onwards the Ministry of Geology and Minerals, then the Bureau of Geology, has been reproducing black and white, coloured infrared and thermal infrared images; reception of Landsat band 4 to (eventually) 7 images began in 1975. The 1:4 000 000 geological map of China was revised with the help of Landsat's 1:4 000 000 mosaic of the country, which revealed a number of previously unknown buried structures and large faults that extend over several hundred kilometres. In 1980 construction started on a ground station near Beijing, to be ready in time to receive, process, archive and disseminate data from Landsat D after its launch. Most recently, Chinese earth scientists have also expressed interest in microwave mappings by the Shuttle Imaging Radar-A. The Institute of Remote Sensing Application, also under the Academy of Sciences, has for nearly a decade been investigating mineral deposits by the application of Landsat data.

The Institute was founded in 1979. Formerly the Department of Aerial Photo Interpretation and the Section of Autocartography, Institute of Geography, it has taken part in remote sensing of seismic geology in northern and southwestern China and organized a series of comprehensive airborne surveys for the purpose of discovering mineral resources in Yunnan province, besides the work involving Landsat data that was mentioned above. It was the first in the country to obtain aerial multispectral photos and object spectrum characteristic data, and to develop an electronic scanning colour separator and image processing system. At present, consisting of five divisions, it undertakes research on object spectrum characteristics of the environment, digital image processing, image analysis and interpretation, autocartography of thematic map, and geo-information system. The general design of autocartographic equipments is its responsibility, and on-going activities include solving hardware and software problems in the integration of the hand tracking digitizer, numerical controlled plotter and automatic Chinese-language typesetter that the Institute itself had originally designed.

A few other institutions in the Academy have work related to geophysical devices. The Chengdu Scientific Instrument Factory (to be described more fully on p 195) has a section devoted to remote-sensing techniques. Items under development there include a field spectro-radiometer, an ultra-low-frequency waveform digital recorder and a dummy colour synthesizer. The Guangzhou Institute of New Geological Techniques was newly established in 1978. It operates four divisions: Remote Sensing, Image

Processing, Holographic Techniques and Electromagnetic Methods for Prospecting. Its current work includes the study of measurement techniques for spectral properties of different terrains, assemblage of a field spectro-radiometer, exploration of the theory and method of seismic holography, and development of a frequency deep-sounding apparatus. The Institute of Geodesy and Geophysics in Wuhan supports research some of which, carried out in the Technical Studies division, have bearing on instrumenta-tion for gravity surveys. Equipment being developed there includes an electronic altimeter, electronic locator, plumb deviator and quartz-sea gravimeter. The other divisions of this Institute in Wuhan are Astro-Geodesy, and Gravity and Earth Tides; the activities of these two divisions will be looked at on pp 184–5. An interesting historical feature of the Institute is that it has existed under its present name previously, from 1961 till 1970, in which year it was reorganized as part of the Earthquake Brigade of Wuhan. In 1978, after the political climate became such that basic research did not need to be hidden under the appearance of applied work and the Academy of Sciences could regain many of those institutes 'lost' to other authorities, it regained independence and reverted to its former designation.

Geochemical prospecting has already been touched upon in Section 4 of the last chapter. In addition to the institutes under the Academy described there, the Institute of Mineral Deposits in the Ministry of Geology and Minerals has sections devoted to radiochemistry and microanalysis, rocks and minerals, and ferrous and nonferrous metals, apart from the division of exploration and prospecting. Another relevant institute is that of Rock and Mineral Analyses. The Ministry's six regional geological institutes all contain Mineral Deposits laboratories, in which mineralogy is studied with microscopy, X-ray diffraction and other techniques.

The Ministry runs six factories in Beijing, Tianjin, Shanghai, Xian, Chengdu and Chongqing which produce hardwares for geological explora-tion. The Ministries of Petroleum, Metallurgical, and Coal Industries each has its own institutes devoted to research on prospecting in its respective field. For instance, at the Central Coal Research Institute in Beijing, efforts in developing modern techniques are at present aimed at combining digital physical exploration, rapid exploratory boring, digital electric prospecting and digital logging.

6.3   Mining Science and Technology

In China four million workers, or about three-quarters of the manpower in the mining industry and nearly eight times that in the oil sector, are engaged in coal acquisition. Although to some extent this number is a reflection of the degree of labour-intensiveness in the coal sector, it remains true that the recovery of any mineral except iron ore has a lower priority. Let us then begin with an examination of Chinese technological standard in extracting this rock fuel.

Coal mining was one of the few heavy industries that were moderately developed before 1949, mostly by foreign investments: the miners were among the first people to experience life in a society that was industrializing in the hard way. Today, the country has the necessary technical knowledge to design fully mechanized pits and to manufacture a range of modern mining equipments. Many mining machinery research institutes exist, an example, from the smaller group that is under direct central control rather than provincial or local jurisdiction, being the institute named after its city of location, Shanghai. On the other hand, manufacturing capacity has consistently suffered from insufficient allocation of capital funds (for a historical perspective of this situation see the first section of the following chapter). Thus, the level of face mechanization among different coalfields or even among pits in the same coalfield can be glaringly non-uniform. Moreover, the highest degree of mechanization is merely about 67 per cent even for the Kailuan mines in Hebei, which lead the country in both annual output and modernization. Under 40 per cent of the national coal acquisition comes from mechanized faces, where the average production per man shift is 14 tonnes. The problem stems, therefore, not so much from any backwardness in expertise as from a privation of long-standing investment in the industry. In the entire interval of thirty years since 1952 the productivity in the coal industry has been raised by a mere 15 per cent, the lowest figure among all industrial ministries; the output of this vital rock fuel has been boosted mainly by requiring more people to join the labour force. Even now, relatively few factories specialize in making coal recovery and transportation equipment. Indeed, mining machinery manufacture has declined in 1980 and 1981 only to enjoy a slight rise of 8 per cent in 1982; by comparison, the mean annual growth rate of light industries over 1979–82 is 14 per cent. In future, the drive towards modernization may be partly fuelled by compensation trade with foreign countries.

The production of mining tools comes sometimes under the Ministry of Coal Industry, Petroleum Industry or Metallurgical Industry when dealing with equipments needed specifically in individual sectors. The majority of manufacturing capacity is administrated, however, by the Third Production Bureau of the former First Ministry of Machine Building, which absorbed in 1982 the Eighth Ministry (Ministry of Agricultural Machinery) and has henceforth been called simply the Ministry of Machine Building Industry. The research arm of this government office is named the Academy of Machinery Research with its headquarters located in Beijing. The state of the art of domestically made extraction machineries may be judged from the specifications of the probably most advanced hydraulic gun, which works at a maximum pressure of 18 MPa and a throughput of 390 cubic metres per hour. Excepting several experimental hydraulic workfaces, however, only three collieries are employing this type of gun. This is despite the fact that the country is one of the three in the world (Canada and the USSR being the others) where conditions permit extensive application of the hydraulic mining method. Pumps for transporting coal slurry to the mine surface typically operates with a 275-metre head at a pumping speed of 800 cubic metres per hour. Current research efforts are being directed towards the development of 'coal-cutters and tunnelling machines, continuous transport facilities, automatic coal lifting, washing and dressing equipments, and computerized communication and dispatch systems'. China made the first set of comprehensive coal-mining machinery in 1978.

Deep mines account for some 95 per cent of national annual coal production. The average shaft depth is 400 metres but the record is 1 060 metres (at Beipiao Coalfield in Liaoning province). One of the drilling machines in common use is the Model 500, a hydraulic type which was designed by Tangsha Institute of Mining Research. In 1978 Shijiazhuang Coal Mining Machinery Plant made drills that can reach 1 800 metres. Retreat long-wall mining is usually employed in dipping coal seams (as in England) but only about a quarter of them are equipped with hydraulic roof supports (unlike in England). For thick coal seams the descending inclined slicing method with caving and the ascending slicing method with hydraulic stowing are often used. Attention at the Central Coal Research Institute (mentioned already at the end of the last chapter) has centred on designing a freezing method for passing through alluvium up to 600 metres thick, shaft lining freeze holes and pre-grouting deep shafts from surface, and adopting road headers for developmental work in coal or mixed strata. Studies are being conducted on the classification of roof strata, under friable and fractured roof and optimum entry layout.

Opencasting, the more efficient method, can be adopted in very few mines, although half of the $1 billion that MACHIMPEX, the Chinese import/export agency, spent in 1979 on capital purchase from abroad for the coal industry was on West German equipment for surface mines. China now manufactures several models of both rotary and rotary-percussion types of drills; electric shovels and trucks or electric railcars are predominantly used to remove overburden and to load coal. Her increase of coal output in the coming few years is expected to come mostly from surface mines. The largest of them, the Fushun Open Pit in Liaoning, is unique in the world in that its overburden, which consists partly of oil shale (p 151), is retained and processed into fuels.

Coal acquisition should not, of course, be the sole concern in coal mines. More 'active research on safety measures' was urged at the 1978 Science Conference. The Ministry of Coal Industry has promised that more research will be supported 'to strive for a fundamental improvement in mine safety'. One of the places noted for such research is the Mine Safety Instrument Development Institute at Fushun. High technology is finding its way into this arena: for example, minicomputers have been installed at Xian Coal Mine and given the function of controlling mine hoists for greater safety. Underground flooding and coaldust explosion constitute the two major hazards. A group at the Central Coal Research Institute is investigating ways to control mine water hazards. Most if not all Chinese mines are gassy: in a quarter of them more than ten cubic centimetres of methane can evolve per tonne of coal per day. At the Central Research Institute efforts have been stepped up on the prediction of coal and gas outbursts, the control of mine fires, and the development of safety testing and mine rescue equipment. The Institute of Applied Chemistry at Changchun (which will be more fully described in sections 5 and 6) recently introduced several gas sensitive semiconductor devices, and explosive-gas detecting systems incorporating them have become available. The Academy of Coal Mine Design in Beijing is responsible for the issue of regulations concerning the maintenance of large mines that are run by the Ministry of Coal Industry. Over a hundred mines have finally set up methane drainage programmes. In these programmes, which require no high technology, drainage conduits are implanted into individual coal seam two to three years before production is scheduled for that particular seam. Moreover, in some places like Fushun bad is turned to good: the drained methane is piped and utilized at low pressure for household heating and cooking.

Furthermore, mining capacity is not just determined by the possession of hardwares. The situation regarding the availability of geotechnical service is again imbalanced. There are many medium-sized pits where little or no local studies on rock mechanics and earth engineering problems are carried out, leading to mining designs that may leave much room for improvement. In contrast, the standard of engineering geology established at specific mines is outstandingly high. The Daye iron ore mine in Hubei is an example. This important openpit work has been threatened by slope stability troubles. The first problem related to the footwall of one area. Its cause was identified as the presence of a premineralization fault of erodable materials and it was overcome by extensive stone pitching. Then, in 1976, on the hanging-wall side of the other end, an 80-metre wide stratum of rock totalling 165 000 cubic metres in volume was observed to be sliding down. With help from the Changsha Institute of Mining and Metallurgy, technical personnel at the mine devised a scheme of remedial measures whereby a total of 132 horizontal anchors were installed. They have mapped the whole region and are maintaining a comprehensive monitor of slope displacements and water levels. Their detailed knowledge of local geotechnical zoning in terms of rock properties and joint and fault orientations and strengths is helping greatly to rationalize a plan of extending and deepening the pit.

Engineering geology is investigated at, among other places, the Institute of Rock and Soil Mechanics under the Academy of Sciences. Formerly the Central-South Institute of Mechanics, this research centre in Wuhan has been involved in the initial analysis of stability at the Daye mine, and also in stability analysis of the Panzhihua opencasting iron mine slope at Dukou in Sichuan province. Work at the Institute of Hydrogeology and Engineering Geology listed in section 1 is chiefly in connection with hydroelectric rather than mining developments.

Basic research on new mining techniques is active in China. As an illustration, a joint programme has been initiated by the Academy's Institute of Microbiology and the Beijing General Mining and Metallurgy Research Institute under Metallurgical Industry Ministry, to investigate mineral extraction by bacteria such as those of the genus Thiobacillus. Practical experience with this method has subsequently been gained in the handling of surface waste minerals. The Xian Institute of Metallurgical Construction is of note in having a department that specializes in mining techniques and machinery for use in small mines. An example of research establishments concerned with ore treatment is the Ministry's Shenyang Institute of Ore

Dressing Machinery. Magnetic gauge, flotation, gravitation, hydraulic and other types of machinery are all under attention.

All in all, the mining sector is for the moment in a state of imbalance. There is sophistication in particular areas but a lack of capital investment in general. There may yet be another problem, and one of socio-political origin. The Chinese press has reported widespread 'private' mining by local authorities or even individuals, that results in serious disruption and waste. Does this problem derive, ultimately, from the over-bureacratization of the power structure there?

6.4 Iron and Steel

'Considerable inter-industry imbalance' serves yet again as the overall description of the Chinese ferrous metallurgical industry. Capacity for the middle stage of crude steel manufacture is strong but those for the two ends, namely raw materials and finished steel production, are weak. Partly as an effort to rationalize this situation and partly as a result of the general cutback on heavy industrial development, it has been planned that from now onwards for a few years major capital investments will be limited to the finished steel sector. Indeed, in the past the annual growth rate of crude steel output has on the whole been decreasing, being about 12 per cent, 8 per cent and 4 per cent in the periods 1955–65, 1965–75 and 1975–80 respectively. Compared to the previous year's figure the production in 1981 actually dropped by 2 per cent, although that in 1982 rose by 2 per cent. China's present-day output of around 35 megatonnes a year ranks fifth in the world.

A common negative factor in the reorganization and expansion of the whole iron and steel industry is, as in the cases of coal and oil, the existence of a transport system that is far from being able to meet all the demands. Bad planning caused by insufficient coordination among different ministries or provincial authorities has not helped. As an illustration, the Panzehua Steel Plant in Sichuan ships all its output downstream along the Yangtze River to Shanghai in the east for finishing, whereas the Meters and Cutting Edge Factory in Chengdu, again in Sichuan, has its special steel sent upstream from Shanghai or by rail from Dalian far away in the northeast. Ore for the Baoshan Steel Work, if and when completed, will be imported through the port at Ningpo that is nearly 350 kilometres away by rail; the Baoshan site, championed by the local authorities and Japanese 'advisers', is not ideal also with respect to geological foundation conditions.

As far as pig iron making is concerned, the most serious problem is that the iron ore and coking coal being mined in the country seldom have good quality. The majority of iron deposits do not make the 35–40 per cent grade. Benefication facilities are fairly modern but wanting in capacity; the types installed include sintering, pelleting and permanent magnet separation. As a consequence huge quantities of scrap and rich iron ore required for blending have been imported from (North) Korea, Australia and Brazil. Such external dependence may decrease in due course because of recent discoveries of major indigenous sources, eg, a 500-megatonne 'rich' deposit in the Jiangyang region of Anhui province. Nevertheless, wherever the sources, the industry will for some time to come still suffer from infrastructural bottlenecks on the ore supply side, as mentioned in the last paragraph. The inadequacy in transportation is already aggravated by the need to move excess quantity of domestic low-grade ore. The distribution of local mines and small iron making plants used to help reduce to some extent the ore movement in terms of volume and distance, but since 1978 more and more of them have been closed for their low efficiencies and uneven product qualities.

The difficulty with coking coal is, fortunately, less intractable. Up to now preparation plants have been built at only four iron and steel complexes to wash raw coal. However, by efforts of intensive research, the Chinese have developed a series of processes that render anthracite and low-grade coking coal suitable for use in the blast furnace. Examples of these methods are selective treatments, rapid preheating and additives mixing. A great deal of attention has also centred on ways of improving the pig iron/coking coal ratio: how to lower coke consumption in iron smelting has long been regarded as an urgent research topic. Slurry coal injection into the blast furnace was a practice introduced in 1970 by the Anshan Iron and Steel Plant, which houses the largest furnace so far made (in 1978) in the country, being of 1 280 cubic metres in working volume. The plant is the subject of a massive programme of modernization, to cost 1 246 million yuan and to be carried out from 1982 to 1985; among the other innovations evolved there is the use of magnesia-alumina bricks. The method of oil injection into the blast furnace is discouraged in China since 1982. The Shoudu (Capital) Iron and Steel Company pioneered the technique of anthracite dust injection and, in 1982, sold this know-how to a British corporation. Another recent achievement of the Shoudu is the technology of the dome combustion hot blast stove, which was acquired by a Luxembourg company through an exchange agreement.

As said before, the country has been more successful in expanding her steelmaking capacity. Sources of difficulties appear to be restricted to the provision of raw materials, including both pig iron and scrap steel that is used to lower the required furnace temperature. Open hearth furnaces still predominate, accounting for 60–70 per cent of the national annual output of crude steel, followed by basic oxygen furnaces 15–25 per cent, side-blown converters 10–15 per cent, and electric furnaces 5–10 per cent. The pattern reflects, on the one hand, historical Soviet influence and, on the other, Chinese reluctance to replace the first type, which, while occupying excessive space, can accept a wider fluctuation in the ratio of steel scrap and molten pig iron charged than the basic oxygen furnace. The most common size of open hearth furnace is of 200 tonnes and the largest, of 500 tonnes. The last two types are built mostly for alloy making. Superalloys for aircraft and other high-technology items are made in electron beam and vacuum arc furnaces. In 1982 China made 3 megatonnes of alloys in over 600 specifications. The technique of electroslag remelting is becoming common after the first block-withdrawing electroslag furnace was fabricated by the Taiyuan Iron and Steel Research Institute in the mid-1970s. Twenty high-smelting point steel furnaces, including vacuum and non-vacuum induction furnaces and electroslag furnaces, have just been added to the steel complex of Daye (cf previous section) at Huangshi city.

A recent major breakthrough has been the construction of a 200-tonne furnace in Shanghai by Beijing Iron and Steel Institute and Shanghai Heavy Machinery Plant. Put into operation in mid-1982, the furnace is turning out ingots for making into forgings, to be used in nuclear power station or as rotars of the 1 000-megawatt steam turbogenerator being designed (see p 155).

Steel finishing capacity is insufficient as is the output of pig iron. Forging facilities are relatively scarce, though the Chinese are rightly proud of those of their 12-kilotonne hydraulic presses that have been developed by their own efforts. For seamless pipes, a diameter of one metre is the maximum so far attained and, furthermore, manufacturing capacity is inadequate: a deal made in December 1982 with Japan was for the purchase of 250 000 tonnes, for example. The steel finished product industry used to cater mainly for heavy industry. Thus, until 1979 only hot rolling mills existed which provided for capital construction purposes, and there was virtually no production of tin- or zinc-coated plates. However, in the last few years, a reoriention towards agriculture and light industries has been effected. The Anshan Iron and Steel Company, the biggest in the country, began recently to make rolled titanium steel for use in bicycles of long durability. In 1979, the new installations at the old Wuhan Iron and Steel Works began operation, though not before December 1981 were they formally commissioned. They involved imported machineries costing US$ 600 millions or half of the total capital layout, though it was found later that some of the items could have been made indigenously. The imports included a continuous casting shop, a 1.7-metre continuous hot mill of 3-megatonne designed capacity with 15 metres per second as the normal rolling speed, and a 5-stand tandem cold mill of 1-megatonne capacity with a rolling speed of 30 metres per second. These mills make tinplate, galvanized plate, hot rolled sheet, cold rolled strip and silicon sheet steel for use in food processing and bicycle, steel furniture, domestic electric appliance and electrical machinery manufacturing industries. Allegedly due to insufficient coordination among the Metallurgical and the Power Ministries and provincial authorities, the mills suffered initially from power shortage that has only just been relieved – see introductory remarks to chapter on Energy. For the present the biggest problem lies with billet supply, which is largely determined by the productivity of the smelters in the old part of the steelworks. Feeds for the smelters come mainly from Daye. The open pit at Daye is undergoing extension and deepening, as said in the previous section, but bottlenecks in ore transport are as much to blame as the inadequacy of mining capacity. The transport situation reportedly continues to be a focus of daily discussion (contention) among the Ministries of Metallurgy, Railways, and Agriculture, Animal Husbandry and Fishery, the last being the other major user of rail transportation.

Most if not all of the major iron and steel making plants have research laboratories attached where R&D are supported. Some centres of higher education have also made contributions to the iron and steel industry. As an example, the Dalian Institute of Technology has developed chromium-nickel-silicon and chromium-nickel-niobium steels that are resistant to corrosions, besides aluminium-silicon eutectic alloys and high Mn-Si bronze among the outcomes from non-ferrous work. Cryogenic properties of low-carbon high-manganese micro-duplex cast steel and the growth behaviour of fatigue crack in 40 Mn 2 A steel are some of the phenomena being currently studied there.

Under the Academy of Sciences, the Institute of Chemical Metallurgy in Beijing was founded in 1958. Initially it has the goals of applying chemical engineering principles and techniques to upgrade the efficiencies of existing metallurgical processes, of proving novel methods of extraction metallurgy and of designing the new equipments thus required. Now, however, its main emphasis has been shifted to chemical reaction engineering and physical chemistry in extraction metallurgy. Among the hundred plus projects it has successfully concluded are: basic study on the intensification of blast furnace operation, designs of the top-fired pebble hot stove and of the top-blown oxygen converter for steelmaking, smeltings of high-vanadium pig iron and of high-phosphorus pig iron, fluidized-bed gaseous reduction for the production of permanent magnet iron powder, and fluidized-bed magnetizing roasting of low-grade iron ore. A high proportion of Chinese ore being of high arsenide sulphide contents, it has taken part in the development of type V106 catalyst, which is resistant to arsenic poisoning.

The Institute of Applied Chemistry at Changchun is another place with active programmes in chemical metallurgy. Its Physical Chemistry division contains the Laboratories of Catalysis, Structural Determination, Optoelectronic Materials, and Electrochemistry, in the last of which metal corrosion and protection are under scrutiny. The Sanming branch of the Fujian Institute of Research on the Structure of Matter has two departments specializing in corrosion and applied coordination catalysis. The main branch is at Fuzhou, where the chief concerns are biological macromolecules and laser materials.

The last unit in the Academy of Sciences system with close connection to the subject of iron and steel is Shenyang Institute of Metals Research. Established formally in 1953, it took as its main task the solution of material problems concerning steel, but later branched out into other areas related to aerospace and nuclear reactor engineering. It is at present organized into 17 scientific and technological divisions. Their designations are, respectively: Chemical analysis; Mechanical testings; Defects and mechanical behaviour of metals and alloys; Structure of crystalline and amorphous states; Structural analysis, fractography and material evaluation; Powder metallurgy and composite materials; Refractory metals, titanium alloys and phase diagrams; Rare earths in steels; Properties of amorphous and liquid metals and of alloys; Superalloys and solidification of alloys; Pyrolytic graphite; Corrosion and protective coatings; Welding and joining; Metal working; Special smelting and high-strength steels; Steels at low temperature and physical properties of materials; and lastly Instrumentation and technical services and information. Among its current applied research programmes on ferrous metals and alloys are the physical chemistry of special smelting, strengthening mechanisms of ultra-high-strength steels, steels at ultra-low temperatures, solidification and crystallization during casting, weldability and plastic deformation behaviour in forming. Research is also conducted in the area of nonferrous metals. Although this subject will be covered in the following section, for the sake of coherence examples of such work are given here. The occurrence of localized slip regions in aluminium-alloys under cyclic stresses is investigated and compared with fatigue phenomena in industrially pure aluminium and aluminium-copper alloys, dislocations in nickel-base alloys during high-temperature creep are examined in the TEM, creep rupture of aluminium is tested through internal friction measurement, experiments are underway on the oxidation of refractory metals at extreme temperatures, and the thermal instability of titanium alloys is studied.

The Institute is well-equipped. Furnace temperature (to 1 900 K) in its creep laboratory and the operation of several servo-hydraulic tensile machines (maximum loading 250 kN) are computer-controlled. Analytical instruments available in it include a Metals Research Quantimet, two SEMs, one of which is a Cambridge S4-10, several TEMs and STEMs (Japanese models), a Chinese-made field-ion microscope, and an Auger spectroscope (Riber). The second of the two scanning electron microscopes was supplied by the Beijing Scientific Instrument Factory. The Factory, set up by the Academy in 1970, made China's first SEM, the DX-1, which had a specification of 100 angstroms in spatial resolution. The one in the Institute is a DX-3 which incorporated a wavelength dispersive X-ray analyser. The DX-5 and a transmission electron microscope are under trial production at the Factory.

6.5 Nonferrous Metals

The Ministry of Metallurgical Industry revealed in 1982 that there were in China 748 nonferrous metal enterprises of various sizes. The focus of this section will be on the production of metals that are of international economic significance.

Traditionally, Chinese expertise in antimony production technology is second to none. Given in addition the present low level of antimony consumption at home and abroad, the low-key development in this field is to be only expected.

Tungsten receives much more attention. Most countries have stockpiles as it is indispensable to the manufacture of, among other things, carbides and high-speed steels which are strategic materials, though molybdenum, of which China also possesses huge reserves of excellent quality, can sometimes serve as a substitute. Chinese reports have come out talking about unspecified but 'significant' innovations in both wolframite and scheelite dressing techniques, and also about improvement in the recovery coefficient of bismuth, which is usually found in association with tungsten. No official figures are available but the annual output of tungsten has apparently declined over the decade, as a substantial proportion goes for export and market price has not been an incentive. On the other hand, should international demand increase, China can no doubt expand production with ease. There are about half a dozen refineries, making each year 500 to 2 000 tonnes of tungsten trioxide 68 per cent concentrate, from ammonium paratungstate as the intermediate. The Xihuashan concentration plant, near a quartz-vein type mine of wolframite/scheelite in southern Jiangxi, has a throughput of 3 000 tonnes per day, and uses a Chinese-designed wet high-intensity magnetic separator to treat sulphidized ore.

The Gejiu region of Yunan currently yields about 60 per cent of the amount of tin mined annually in China. Nine concentrators and a central smelter are located there. The ore is chiefly lode tin, often of high grade – the average concentration being 5 to 6 per cent but containing much base metal impurities and so causing major beneficiation and metallurgical problems. Oxidized ore is usually treated by gravity concentration. Sulphide ore is crushed and ground in ball and rod mills, then floated to separate out a copper concentrate. The trailings are subsequently treated on tables. Smelting, both primary and secondary, is done in coal-fired reverberatory furnaces. Refining methods employed include drossing, poling, aluminizing, liquation and chloride solution electrolysis. The national number two tin centre is the Hichih district in the neighbouring province of Guangxi. A new mine/mill complex is being built there to produce tin, antimony, zinc and lead, and other related elements.

A fairly comprehensive extraction, separation and application industry for rare earth elements now flourishes in the country. The Bayan Obo iron mine (which we have come across in Section 1) in Inner Mongolia was the first to be tapped and is still the biggest source. The Chinese have made a series of contributions to rare earth technology, one being the optimization theory of cascade separation processes that was developed at the Chemistry Department of Beijing University in late 1960s and early 1970s. Separation is currently done through ion exchange or with complexing agents. There are now twenty refineries in the country with the city of Baotou boasting seven. The research department of Baotou Iron and Steel Company has been prominent in the work on ore dressing and refining for many years. The Company was the host to the 1975 National Rare Earth Conference, and now produces the lanthanides in 100 specifications as concentrates, pure oxides or metals. It also sells 18 types of steel magnetic alloys covering a range of tensile strengths and shapes. Nationwide, altogether 20 kinds are in production, which include low-alloy and spring and high-speed steels. Changchun Institute of Applied Chemistry, which will be more fully described towards the end of this section, is outstanding for its work on the syntheses of rare-earth compounds. The application of rare earths as converters of soft X-rays or ultraviolet radiation into visible light in medical and optical apparatuses is under scrutiny at the Chemistry Department of Fudan University in Shanghai.

In recent years China has stepped up the output of titanium, which is an essential material for the aircraft and missile construction industry. One of the key centres of this industry is Shenyang, near which city, at Fushun, is a plant which produces annually 500 tonnes of titanium sponge using the

Kroll method. In 1980 a concentration plant went into operation in Sichuan province. Embodying hitherto the largest project to exploit vanadium/titanium/magnetite ore, the Sichuan plant has been built mainly with Chinese technology and is designed to provide each year 50 000 tonnes of titanium concentrate and 6 000 tonnes of sulphur-cobalt. China has made important innovations: in 1979 a new, more efficient method for separating titanium from iron was successfully tested. Consideration is also being given to the erection nearby of a titania pigment factory of 300 000-tonne annual capacity. The Chinese are moderately advanced in the associated technology. They have independently perfected the plasma technique for preparing titanium white to replace the sulphuric acid process (see below, in the paragraph describing the Institute of Chemical Metallurgy). However, the titanium trichloride plant completed in 1979 near Beijing was purchased from Toho Titanium of Japan.

China is well-endowed in many rare metal resources – the reserves in niobium/tantalum may rank first in the world – and development is also emphasized in this sector. At Hangzhou towards the end of 1982 an international conference on 'Production and Application of Less Common Metals' was held, during which the host country was strongly represented.

The more commonly met representatives among those nonferrous metals in which China remains a large-scale net importer will now be looked at. The first sector to be examined is that of aluminium which, like titanium, carries great importance by virtue of its role in the aerospace industry, besides the energy and other industries. (Its use in the motorcar industry is unlikely to become significant in proportion, as China quite rightly has not put any priority on city car manufacture, but more will be committed to rail transport, as real efforts are underway to reduce the unit weight of rolling stock: see end of Chapter 8.1.) The geological reserves of aluminium are enormous but its workable resources are small, and moreover the current level of production is far from adequate. The level has been almost static at around the 358 000-tonne mark over the last few years, while consumption is roughly half as much again and rising slowly but steadily, being 510, 550 and 560 thousand tonnes in 1977, 1980 and 1981 respectively. The biggest reduction plant is at Fushun (which, as noted above, is also a titanium centre), producing each year 100 000 tonnes from locally mined aluminous shale by the use of horizontal stud Soderberg cells. Smaller facilities exist in Changtien, Zhengzhou, Sanmen Gorge, Yiling and Baotou. At Guiyang a plant has just gone into operation, making use of electricity supplied by hydropower stations on the Wu and Maotiao Rivers (see pp 156 and 157–8). A new major installation may emerge in association with a huge high-grade bauxite deposit recently located at Pingguo in Guangxi province. A constraint on production is the generally large proportion of insoluble silica in Chinese ore, which is therefore unsuitable for the Bayer process. Another limiting factor is the severe demand on electricity at the electrolytic reduction stage – aluminum refineries have earned the nickname of 'electricity tigers' in China – though recent technical innovations are said to have brought about greater efficiencies. According to a press account in 1982, the Fushun Reduction Plant, in cooperation with Liaoning University, completed the development of a microprocessor controlling system by means of which electricity consumption could be 'reduced by 1–2 per cent'. China exports the metal with purity specification up to 5–9 (99.999 per cent). She was also the first to use the technique of casting and rolling aluminium sheet directly from the melt.

The country turns out annually around 300 000 tonnes of copper, an amount that falls far short of domestic needs as in the case of aluminium. Imports annually account for half or more of total nonferrous metals purchase, having been bought from the UK, Japan, the Philippines and Chile (in the last instance through long-term contracts signed before the change in Chilean government). Shipments have also come from Peru in return for Chinese help with the Tintaya copper mining project. From Zambia 50 000 tonnes are due each year, as part of the repayment to the interest-free loan to build the Tanzam railway, concerning which China has given generous technical help as well: however, repayment has been allowed to be rescheduled repeatedly and in January 1983 was forgone completely as a gesture of friendship. Nevertheless, the supply situation may improve before long, if the expansion plan envisaging a large nickel/copper open pit at Jinchuan in western Gansu is implemented without delays, and when the porphyry orebody extraction at Dexing in Jiangxi province (an operation which can eventually provide 400 000 tonnes of contained copper annually plus such byproducts as molybdenum gold and silver) goes ahead. An integrated smelter is under construction at Guixi, Jiangxi, to treat ore from Dexing and elsewhere. At the moment, the main producers for the metal are Daye in Hubei province, Kuming in Yunnan and Tongling in Anhui.

At Daye (pp 130 and 133) iron is mined to be fed to the Wuhan Iron and Steel Plant but significant copper contents are present. The Daye Nonferrous Metal Company, located near Tonglushan, operates seven smelting and chemical processing plants. An integrated electricity generating station that makes use of the waste heat from smelting in the reverberatory copper furnace entered service in 1982 (cf introductory section, next chapter). The Kuming complex comprises five mines, served by a central smelter which treats concentrates averaging 20 per cent copper, 10 per cent iron, up to 50 per cent silica and 10 per cent sulphur. There the concentrates are pelletized and indurated on a travelling grate drier before being smelted in one of two Soderberg electrode matte furnaces, to become 50–60 per cent copper. At Tongling, where the deposits consist of skarn copper, a new smelter-refinery has just come on stream. Its feedstocks come from the surrounding mines as well as places further away. At the local concentrators the mine ore, of 0.5–1 per cent grade, passes through two-stage cone crushers and closed-circuit ball mills; subsequent flotation is carried out using a rougher, two cleaner and two scavenger stages. The concentrates thus obtained assay 22–24 per cent copper and recovery is rated at 93–96 per cent. After arriving at Tongling in railcars the concentrates undergo three-stage briquetting, then smelting in an open-top blast furnace and two converters, followed finally by refining. The old smelter had no electrolytic facilities, so that the blister copper used to have to be transported over 500 kilometres of railways to a refinery in Shanghai. Worse, it was a notorious environmental pollutor because of low sulphur dioxide recovery from blast furnace and converter gases. At the new smelter, an average of 5.5 tonnes of sulphuric acid appears as a byproduct for every tonne of blister copper. The Shanghai refinery remains the largest in the country and there emphasis is on retaining gold, silver, selenium, platinum and palladium in the anode slimes, while treating copper concentrates and scraps.

Some copper also come from the multimetal smelters at Zhuzhou in Hunan and at Shenyang, which deal with ores of copper, zinc and lead but in different proportions. The main zinc/lead centre is, however, at Fankow in Guangdong, where there are proved reserves of 40 megatonnes of 11 per cent zinc, 5 per cent lead and 20 per cent sulphur (existing as galena, sphalerite and pyrite) mineralization. Unfortunately, the concentrators at Fankow cannot achieve high throughputs and have an aggregate capacity of merely 3 000 tonnes per day. The poor performance is attributed to difficulties in treatment caused by the prevalence of fine-grain intergrowth between the three sulphide minerals, necessitating excessive grinding and flotation. The output is sent to the zinc/lead blast furnace at neighbouring Shaoguan. Three types of concentrates comprise the feedstock: 'zinc' (45-per cent zinc 2-per cent lead), 'lead' (5-per cent zinc 40-per cent lead) and 'mixed' (28-per cent zinc 11-per cent lead). Appreciable amounts of copper, silver (contained in galena), cadmium (in sphalerite), mercury and arsenic also occur. The concentrates are first sintered on an updraught sinter machine; the sulphur dioxide fume so evolved is removed in a single absorption sulphuric acid plant. Sinter, together with metallurgical coke, is then fed to the furnace, which is of the Imperial Smelting Process type. Lead bullion is treated by electrolytic refining and zinc metal, after being cast into plates, is treated in a refluxer plant employing silicon carbide trays. Besides Fankow-Shaoguang, there are smaller zinc/lead bases in the adjacent provinces of Guangxi, Hunan and Fujian and in Liaoning in the northeast; the new Hichih mining and refinery complex in Guangxi has earlier been mentioned. The total annual production is in the region of 300 000 tonnes each, a figure near to that for copper and, likewise, short of the respective levels of demand. Zinc and lead have been imported from countries such as the UK, Japan, Peru and (North) Korea. It may be interesting to end this overview by noting that, historically, Chinese miners had for many centuries worked lead-silver ores in what are today Vietnam and northern Burma, using techniques which were at that time the most advanced in the world.

The Chemistry Department of Beijing University, in cooperation with the Laboratory of Physical and Chemical Analyses in the Institute of Physics and with Beijing Institute of Rare Earths, has made composition determinations of a series of technologically important materials and excels in the micro-analysis of metals such as indium, tellurium, strontium, scandium, yttrium, vanadium and the rare earths. The Department has good facilities for chemical analysis, and the Laboratory in the Institute of Physics is well endowed in physical analytical instruments. They include ion-selective electrode apparatus, X-ray fluorescence spectrometer, and several atomic absorption spectrometers one of which is a Perkin-Elmer Model 703 with an HGA 500 graphite furnace.

The Institute of Applied Chemistry at Changchun, also belonging to the Academy of Sciences, has three research centres with a total of sixteen laboratories. The first centre is concerned exclusively with high polymer science, but one result from there has been the application of rare earths in complexes as catalysts for the stereo-specific polymerization of diolefins. The Inorganic and Analytical Chemistry Centre has been outstanding for its work on rare earths. It has systematically devised means for the extraction, isolation, separation, purification, trace analysis and structural determination of fifteen rare earth elements. Much has also been achieved regarding the syntheses of lanthanide compounds and preparations of alloys, particularly materials for semiconductor lasers (Chapter 8.6). Theoretical study on the fundamental properties of rare earth compounds has also been carried out. Activities at the third centre, entitled Physical Chemistry, have already been discussed in the last section.

The Institute of Chemical Metallurgy, mentioned previously on p 134, is composed of seven laboratories devoted respectively to: Pyrometallurgy, Hydrometallurgy, Fluidization, Computation, Chemical analysis, Physical methods of examination, and Instrumentation. Notable examples of past investigations conducted at the Institute are on the extraction of chromium and vanadium and their separation through primary amines, and blast-furnace smelting of vanadium-bearing titaniferous iron ore. The novel way of producing pigment-grade titanox through vapour-phase oxidation of titanium tetrachloride was developed there from 1974 to 1976. Employing megahertz-range plasma heating, the technique attained a 98 per cent yield of titanium white of granularity less than 0.4 microns, while reducing the excess oxygen coefficient to under 1.1 degree per kilogram. In the last few years the Institute has worked on pneumatically controlled multistage fluidized bed, high-pressure hydrometallurgical process for the treatment of oxidic copper ore, carbonated leaching of vanadium mechanism and equipment, high-pressure hydrogen reduction for the preparation of powdered metals, and the treatment of electrolytic slimes of nickel-based alloys by reduction and ammonium leaching. A recent development is the study of metallurgical processes via computer simulation.

An on-going project at the General Research Institute for Nonferrous Metals in Beijing is aimed at discovering ways to improve the performance of concentrators at the zinc/lead mines of Fankow (see above). The Institute owns some unusual equipment, such as a Chinese-built superconducting magnet separator that incorporates niobium-titanium alloy superconducting wires manufactured in China, and a Japanese-made 1-megavolt high voltage electron microscope for the observation of thick specimens. It pioneered the use of neutron activation techniques in the early 1970s and has applied them to metals of high purity grade and to semiconducting materials.

7 Energy Science and Technology

The Chinese people are known to have mined coal over a thousand years ago – for making ink, among other purposes. Several ancient texts mention 'weak water'. described as upon which even a boat made of feather would barely float; it is they, rather than the Old Testaments as is often supposed, that contain the earliest records of petroleum seepage in the world. Later passages dating to 2 300 years ago explained in details the use of 'oil water' for greasing cart axles and, in one place, for combustion. The year 211BC can be pinpointed as the time when a natural gas well was drilled with bamboo poles in what is the Sichuan basin. Possibly this gas well was not the first to be sunk there, as drilling with bamboo poles had been going on for centuries beforehand: in 600BC Confucius mentioned drilling for salt in this way. Watermills representing the way of tapping hydraulic energy that eventually evolved into hydroelectric generation, appeared in the first century AD or perhaps some time before. Much earlier still, in a part of the 'Book of Changes' that probably had survived from the end of the first millennium BC, reference was made to 'fire in the damp' meaning evidently marsh gas. Around AD630 a vertical-axis windmill appeared which from its descriptions in contemporary writings qualified as the embryonic form of the modern wind turbine. Back in 700BC, the existence of subterranean hot water was discovered although not until the third century BC, during the reign of the first emperor of united China, was the first hot spring bathhouse built at what became the 'Huaqing Pool', and the site of a famous historical incident two millenia later. The utilization of subterranean hot water in processing silkworm cocoons and silk was a widespread practice by the third century AD.

Today, as should perhaps be expected of a nation possessing the third largest land area, but inappropriate to the most populous country in world ranking, China is placed third alike in total resources and production of primary energy. Broken down into different sources, the reserves are quantified in Table 4 of Appendix 1 at the end of the book. The relative importance of individual sources is summarized in the following Table 5.

The expansion of the energy industry has been impressive, if we take its pathetic state in 1949 as the baseline. Nevertheless, the energy (and transport) sectors constitute the weakest links in the infrastructure that underlies economic growth on all fronts. In the Chairman's Report to the Twelfth Party Congress convened at the beginning of September in 1982, 'energy shortage and the strain on transport' were singled out as the 'major overall checks on China's economic development at present'. (Chinese railways which play the primary role in fuel transportation will be discussed in the first section of the following chapter.) Writing in a 1983 issue of the 'Beijing Review', a staff member of the Economic Research Centre under the State Council explained that 'energy, transport and technical transformation of existing enterprises are the key links to developing China's national economy'. Per capita energy consumption has been able to increase twentyfold since 1949 but, in absolute terms, has reached merely the level of 800 kilograms of coal equivalent (kgce). This level is about quadruple those of India and other 'lower income countries' but only a third of the world average. More pertinently, the growth rate of energy supply fell consistently behind that of gross industrial output, which, in mean annual figures, were 9 per cent and 10 per cent respectively over the years 1952–78. At present around 20 per cent of industrial manufacturing capacity is estimated to be idle as a direct consequence of energy shortage, which itself is mostly caused by infrastructure difficulties in fuel transportation. The situation in electricity generation is the most serious. In many cities factories are shut on different days of the week so as to reduce the peak load on local grids. For three years the imported 1.7-metre rolling mills of Wuhan Steelworks (section 4, previous chapter) was never operated to full capacity because of uncoordinated planning that led to chronic power shortage. This particular difficulty was not alleviated until the completion of the first stage of the Gezhouba hydroelectric station (described below in section 3).

Besides some government ministries that cooperate poorly because of poor lateral communication, the end-users themselves are partly to be blamed for their common energy inefficiencies. The prices of fuels and coal in particular are subsidized by the state so that those factories that are solely concerned with fulfilling production quota have little incentive to invest in newer, more efficient equipment. It is reckoned that the mean calorimetry is 58 per cent for China's 180 000 industrial boilers, 30 per cent for thermal power plants, and 25 per cent for industrial furnaces and kilns; the corresponding figures in the West are 80 per cent, 35–40 per cent and 50–60 per cent. In Chinese steelworks an average of 1 200 kgce are needed to refine one tonne of steel, half as much again as in the USA. Indeed, how to lower coke consumption in iron smelting and energy consumption in energy-intensive metallurgical and chemical industries has been identified as a vital research problem (sections 4 and 5 of the last chapter).

Already, a conservation programme has been embarked upon and met with considerable success. Gross industrial output was up 8.5 per cent in 1979, 7.2 per cent in 1980 and 4.1 per cent in 1981 against the respective preceeding year, while energy consumption rose by 2.5 per cent, 2.9 per cent and 1 per cent correspondingly. The government allocated 3 000 million yuan for conservation measures in 1981, representing a 50 per cent increase over the budget of the previous year. In 1981, energy consumed per unit output dropped by a further 6 per cent. In comparison, the overall elasticity coefficient of energy to output was atrociously high at 1.7 throughout the period of 1952–80 (1.4 for the seventies). The improvement is due partly to the shift from heavy to less energy-intensive light industry during 1979–81, and also partly to achievements in conservation. As an example, after technical transformation, the Capital Iron and Steel Company in Beijing consumes on average 990 kgce to produce one tonne of steel, a reduction of 250 kgce. Technological advances have been made especially in the exploitation of waste heat, which also serve to reduce thermal pollution. Since 1981, integrated electricity and heat turbogenerators have been installed, many of them at small chemical fertilizer factories. The biggest of such power stations is that run by the Daye Nonferrous Metal Company. With exhaust heat from a copper furnace as intake, the station consists of a boiler and two generators giving a combined capacity of 9 megawatts.

In the immediate future, the power shortage may be relieved by imposing rationalizing and conservation measures and in the medium term, by investing more in the exploration for fuels and the construction of energy plants. During the period of the Sixth Five-Year Plan that ends in 1985, the government will earmark just under 60 000 million yuan, a quarter of the total budget for capital construction, to the development of energy. Ultimately, however, the controlling factor in the energy equation will lie in technical competence regarding fuel extraction, processing and transportation, as well as power generation and distribution. This chapter seeks to provide a perspective for Chinese technological expertise, under the headings of various forms of energy: in the 1978 science plan 'energy' has been placed second among the eight specified research priorities.

7.1 Coal

In the near future China is bound to redouble her efforts to tap coal and hydropower resources. The relative contribution of oil is set to decrease, an event that may be associated with the apparently declining influence of the Petroleum Ministry in central policy making. Hugh oil reserves exist but their impact on the production figures can only be viewed in the medium- to long-term perspective. Coal output, on the other hand, is likely to rise at 1.8–3 per cent each year in the period of 1982 to 1987, thus sustaining an average 1–2 per cent increase in overall energy production despite oil losses. The hope is for this rise to accelerate; in the development programme announced by the Ministry of Coal Industry in October 1982, annual production of the rock fuel is planned to be doubled by the end of the century.

Coal reserves are present in adequate quantity and satisfactory in general quality, with the majority in the category of hard coal, 17 per cent being anthracite, 37 per cent coking coal, 32 per cent steam coal, 8 per cent lignite and 6 per cent subbituminous. Of economic reserves, 87 per cent can be mined with the use of underground methods and the rest with surface techniques. As already discussed in Section 3 of the last chapter, the Chinese have acquired the necessary expertise to design fully mechanized pits and to manufacture a range of modern mining equipment. Their actual capacity of coal extraction (and transportation) constitute 'a weak link in the national economy' and has been so for the last two decades. Since the sharp fall following the Great Leap Forward era, the annual output figure of 280 megatonnes in 1960 was not regained until 1969. Raw coal supply rose spectacularly again in 1969–72 but in fact, by 1973, it could no longer meet the demand for anthracite and coking coal, so that nearly half of the country's small nitrogenous fertilizer plants had to be switched to using coal of lower quality, that could be obtained locally. Around the same time, wholesale conversion of industrial installations from coal to oil burning began. Oil-fired boilers are cheaper to maintain since they do not produce ash; another reason for the conversion is that, reportedly, the 'Gang of Four' then in power wished to force a reduction of petroleum export 'in the interest of resources sovereignty'. The present change in the reverse direction – see next section – is therefore easier than might be feared. Indeed, from Table 6 of Appendix 1 we find that from 1949 to 1960 coal accounted for over 95 per cent of national primary energy production; in 1974 its share had slipped to two-thirds and up to now the proportion has remained below three-quarters. Moreover, despite its high export potential, to Japan for example, it has earned much less foreign exchange than petroleum: in 1980 only 6.2 megatonnes were shipped abroad, to (North) Korea, Romania, Bangladesh and Japan. However, the National Coal Import and Export Corporation has embarked upon an export drive, and the National Coal Development Corporation was set up in July 1982 to step up the modernization of existing mines and construction of new pits. Very recently a deal was struck with Occidental Petroleum Corporation, regarding a joint venture to develop the 1.4-gigatonne deposit of high-grade steam coal located at Pingshuo in Shanxi province.

The downward trend reflected not so much the growth of oil production than a persistent underinvestment in the coal sector itself since the 1960s, though the two phenomena could have been politically related. The sector received nearly 12 per cent of funds earmarked for all industries under the 1953–7 Five Year Plan, during which time both output and productivity climbed rapidly. At the end of the succeeding three-year period (the Great Leap Forward) as mentioned above there was a sharp fall. One of the reasons was a virtual halt in the opening of new coalfields. Since then, and not until the last few years, capital construction in this sector has received low priority.

The dramatic increase in coal output during the Great Leap Forward has been partly due to the development of native pits. To this day, local mines are still encouraged and helped by reduced taxation as well as by being allowed to sell coal at slightly higher prices. They now number over 20 000 and operate in 1 221 of China's 2 137 counties. In comparison, there are just under 1 000 state-owned mines which are administrated centrally by the Ministry of Coal Industry. Chinese reliance on both types of mines is in accordance with the policy of 'walking on two legs'. In this particular case, the small-scale mining serves to mitigate the traffic congestion caused by large-scale coal transportation, to diversify the local economy and, with strategic considerations in mind, to prevent the over-concentration of resources. Unfortunately, because of their nature, small pits are in general labour-intensive and poorly mechanized. Overall, in terms of output less than 40 per cent of coal extraction in the country is mechanized. Nevertheless, the Ministry of Coal Industry said in 1982 that it will 'continue technical transformation of major mines in a planned way and achieve the goal of 56 per cent mechanized operation by the end of 1990'.

The coal preparation segment is also starved of funds. Even a large and modern mine like the Fangezhuang Colliery, which produces 3.6 megatonnes a year, has no integrated washery but merely dry sorts. Only one sixth of the national output is treated (and used predominantly for metallurgical purposes), at 17 central plants and the 99 plants that are located in specific coalfields. Over half of these plants have annual capacities below 1.5 megatonnes. (There are also four attached to particular iron smelters.) Individual plants may attain technical distinction. The one at Zhuzhou, Hunan, was cited as the national model for 1982: it recorded a washing efficiency of 95 per cent and the highest grade of coal prepared had an ash content of 4.6 per cent or less. On the other hand, the expansion of preparation capacity has been lagging behind that of production, amounting to 2.65 and 12.43 megatonnes respectively in 1981 for instance. The 1982 Plan of the Ministry of Coal Industry envisages, however, that by 1990 the proportion of washed coal will be raised to 56 per cent.

At present coal cleaning is mostly done by jigging. The heavy media separation process, being more efficient but more expensive, is less commonly employed. For new installations, 70 per cent are planned to use jigs and froth flotation. Flotation cells of up to 10 cubic metres in unit volume and 1.2 tonnes per hour-cubic metre in unit capacity are made. The first jet turbulent flow flotation assembly, designed and built by the Nanshan Coal Dressing Plant, is claimed to be more efficient and easier to maintain than many foreign models. The Institute of Coal Dressing Design in Beijing is among the other outstanding organizations that undertake research into preparation techniques. A small unit at the Central Coal Research Institute (p 129) is studying dewatering methods, filtering and deep cone thickening included.

Coking coal in China generally has good mechanical strength and, contrary to the impression of many Westerners, a low sulphur content of 0.5–1 per cent. It does, however, have medium to high ash content of 7–35 per cent, mostly in the form of silica and alumina, with ferric oxide usually constituting under 5 per cent of the ash. Hence washing is indispensable before its utilization in the blast furnace. The largest plant so far constructed for this purpose is at Pingdingshan in Henan. There 3.5 megatonnes are processed each year, by inclined wheel heavy media separators for the 13 to 300 millimetre fraction, heavy media cyclones for the 0.5 to 13 millimetre sizes and froth flotation for the minus-0.5 millimetre fraction. The Coal Preparation Design and Research Institute, also located there, is looking into the improvement and reduction in operation cost of resonance screens, which do the jobs of preliminary screening, desliming, dewatering, draining and rinsing. Elsewhere, the cheaper, albeit less efficient, vibration screens are almost universally installed. A new coking coal preparation plant of 4-megatonne annual capacity comes on stream in 1983. Some of the equipment is West German import, but others are of indigenous manufacture.

Steam coal amounting to 110 megatonnes was consumed during 1979 for electricity generation. Containing 0.3 to 1.9 per cent sulphur and 5–40 per cent ash, it was usually burnt without being washed first. Research and development efforts are being put into its preparation procedures, the motivation partly deriving from its export potential. It has been earlier mentioned that a gigantic field in Shanxi (of which more below) will be opened with foreign participation.

Bottlenecks in transport constitute perhaps the most severe limiting factor in Chinese utilization of coal, the help from scattered but small mines notwithstanding. Almost 95 per cent of reserves is found in the north, with over a third concentrated in Shanxi. This province currently provides nearly a fifth of the coal extracted in the entire country. There, tunnel and slope mining and sometimes openpit casting are practical under the local geological conditions: the overburden is seldom great and deposits exhibit few geological complexities (folding and faulting). In view of these advantages, a coke producing centre in its central region and an anthracite centre in its southeast have been selected as sites of major development. On the other hand, because of insufficient transport facilities – despite the interesting anecdote that one trainload of coal leaves it for other provinces every five minutes – coal stockpiling in the whole province exceeded 37 megatonnes by late 1982 and is expected to reach 47 megatonnes before the end of 1983.

Indeed, all the twelve leading coal mining centres that have outputs on the 10-megatonne level are situated north of the Yangtze, and the necessity to 'move north coal to the south' on large scale is a problem that has not been fundamentally eliminated, despite the discoveries of several large coalfields in southern provinces. The Southwest China Natural Resources Exploitation Corporation was set up in 1982 to coordinate development policies of different government agencies and to attract overseas capital, particularly for the operation of coal mines in the provinces of Hunan, Guizhou, Yunnan and Guangxi. Guizhou leads in proved reserves (46 gigatonnes) and it may well be the location of the largest and richest steam coal deposit in the world. Concerning it preliminary agreements have just been reached with a consortium headed by United Development Incorporated and composed of companies from (West) Germany, France, Belgium, Spain and possibly also the UK, all of which will accept coal in payment for their costs. The agreement forsees the opening of 29 new pits, upgrading of three existing ones as well as a gigantic infrastructure project that combines construction of a power plant and a rail link with port modernization at Zhanjiang in Guangdong to handle the export. The whole scheme represents the biggest undertaking currently being contemplated by China at present. The Chinese Corporation includes representatives from the Ministries of Coal Industry, Railways and Communications, to faciliate lateral communication and coordination. Progress will not be immediate, however.

At present coal accounts for around a third of the total tonnage hauled on the railway system. At the 1980 Mine Planning and Development Symposium held in Beijing, officials pledged that, to improve the situation more rail links and coal freighters were to be built, some lines reconstructed, ports expanded and new ones fabricated. With the aim of reducing the load on railways – and the pollution of urban air – large thermal power stations with an aggregate capacity of 10 gigawatts are under construction at several pit mouths. High priority has also been allocated to research on the use of low-quality deposits, which are typical of coalfields in the south. An integrated transport development project for Shanxi, started in April 1982, will bring about the electrification and double-tracking of the Datong to Fengtai (on the outskirts of Beijing) line, the double-tracking of the one between Taiyuan and Datong, and that of the Taiyuan to Jiaozhou railway. Already open to traffic, since September 1982, is the 235-kilometre link from Taiyuan via Yangquan (a major anthracite-producing centre) to Shijiazhuang; this line is both electrified and double-track, the first to be so in the country. According to projections, with the aid of this facility 30 megatonnes of coal can be moved out of Shanxi annually. Support is also being given to research efforts to render coal less inconvenient to transport than oil and gas. A short slurry pipeline is undergoing trial in Guangdong, and a feasibility study has been completed on the laying of such pipelines from northern Shanxi to the port at Tianjin.

Till now most of the export coal is shipped from Ginhuangdao and the rest from the ports of Qingdao, Shanghai and Nanjing. Besides the pipeline planned for Tianjin and the port modernization for Zhangjiang, an expansion project at Qinhuangdao has already moved forward. A port specializing in loading coal is scheduled to be completed there in 1985 to handle 20 megatonnes each year. An electrified line linking Qinghuangdao with Beijing will enable coal to be moved from Datong via Beijing easily. It may be noted that all the aforesaid places, namely Qinghuangdao, Tianjin, Qingdao, Shanghai and Nanjing, also comprise – with the addition of Dailian (formerly Lu-da) as well as Xianggang (Hong Kong) – all of the existing outlets for oil export.

In a few areas of coal technology China is by no means backward. An example is coal gasification. The Central Coal Research Institute has made many advances with its Lurgi pressure unit, and work of a similar nature has just started at the Taiyuan and the Yantai Institutes of Coal Chemistry. The country's first full-scale lignite pressurized gasification plant began operation in mid-1983 at Shenyang to produce 540 000 cubic metres daily. Under the Ministry of Petroleum Industry, the Dalian Institute of Petroleum Research (of which more in the next section) is also looking at gasification using steam and oxygen under high pressure. Under the same Ministry, the Beijing Institute of Petroleum has reported significant progress concerning low-temperature carbonization of pulverized bituminous coals in fluidized bed. Another area in which China has kept up with the West is the liquefaction of lignite. The Liquid Fuel Synthesis group in the Dalian Institute of Chemical Physics is one of the bodies of researchers that have been engaged in long-term work on hydrogenation liquefaction. In 1981 a Chinese press account claimed that test had been completed on a process method based on solvent extraction, by which conversion efficiency approaching 88 per cent was realized.

Coal energy conversion, including gasification and liquefaction, is the newest growth area of research at the important Shanxi Institute of Coal Chemistry under the Academy of Sciences. Originating from 1954, this Institute has, in the past, successfully completed a number of major projects, such as the classification of Chinese coking coal and the use of high-volatile bituminous coal in coke ovens. A recent press release reports that the Zhejiang Institute of Coal Research, a provincially run unit, has achieved 'significant progress in the comprehensive utilization of stone coal'. The multipurpose utilization of different grades of coal has been designated at the 1978 Science Conference as a key focus for research. In a decade's time, after a period of sustained investment, with technological break-throughs such as the liquification and gasification of coal, and stimulation by the worldwide 'rediscovery' of this form of fossil fuel, the coal industry can look forward to national and international prominence.

7.2 Oil and Gas

Oil is important to China, being required not only in large quantities by domestic users but also as a primary foreign exchange earner. Before 1949 the situation was just the reverse. She consumed little oil, amounting in the entire first half of the century to only about 30 megatonnes of her own production in 1970, and over 90 per cent of the amount was imported. In those days the country, having predominantly lacustrine sedimentary basins, was regarded abroad as 'oil poor' (section 1 of last chapter). The accusation has been made that such an annulment of indigenous resources was encouraged for the aim of preserving the market for American petroleum products and particularly kerosene.

After 1949, accompanying the rapid industrialization, oil consumption rose fast but home production failed to keep pace. In 1958 about 7 megatonnes were consumed, nearly half of which was purchased from the Soviet Union. One consequence of the worsening of Sino-Russian relations was, therefore, a sharp deterioration of supply of hydrocarbon fuels: for a while buses were running on coal (personal memory). Another was the threat of a disruption in the exploration programmes in the north and northeast, programmes which has been going on for some years with the participation of the Russians, who wanted to build a substantial petroleum base in these areas to serve 'their' own Amur-Ussiri industrial regions. In the spirit of self-reliance then catalysed by self-preservation, workers and geologists endured enormous hardship not merely to continue but also to speed up the prospecting and test drilling. The victory of this all-out effort was symbolized by the striking of oil in the northeast at what became Daqing Field, on 10 September 1959 by a team led by the late 'Iron Man' Wang Jinxi who was subsequently hailed as a national hero. The develop-ment of this extensive oilfield helped to overcome the oil crisis, restore economic sovereignty and enhance the credibility of the self-reliant prin-ciple. The close of the Great Leap Forward era marked the beginning of the shift towards basic self-sufficiency. Before 1963 there was only import and no export of crude oil; after 1969 there was no net import. Diesel oil and petrol have never been purchased from abroad again since 1969.

The performance of the oil industry has been astounding, both in absolute terms and in comparison to any other sector of the economy, but problems have arisen in the last few years. Since 1973 when oil was distributed to Xianggang as well as sold to neighbours like Japan and Thailand and recently to North America, the government has been relying more and more on oil revenue, which stood for example in 1977 at US$1 000 million or 13 per cent of total export earnings and in 1981 at the same percentage. From the late 1970s onwards, however, the growth of oil production has once more failed to keep up with home demand. Sectors responsible for the fast increase in consumption have been rail transportation (dieselization), agriculture (tractors and pumps) as well as chemical fertilizer and synthetic textile manufacturing (both to boost grain harvest, the former by raising soil productivity and the latter by freeing land committed to cotton cultivation). In addition, since around 1973 many electric thermogenerating plants were converted from coal to oil burning, because of the fall in coal output. On the other hand, because of priority shifts, for several years at the end of the 1970s internal funding for exploration activities was reined in. An oil industry investment budget of 4 500 million yuan in 1978 fell by a whole 20 per cent the next year. Indeed, as may be seen from Table 7 in the Appendices, oil production has actually been dropping since 1980, mainly due to the faltering output from the old (depleted) Daqing Field that accounts for about half of the national total. Annual production is going to fluctuate just above the 100-megatonne mark or more probably to decrease slightly but steadily by up to 5 per cent a year, until offshore oilfields come on stream (with the help of external investments) in 1986–88, by which time output should scarcely satisfy domestic demand if the latter is to suffer no harsh squeeze, leading little room for exports. The government has announced the intention of encouraging industries to replace oil by coal, when consumed as fuels, as much as possible by 1990. Annual output then is envisaged to be 50 megatonne onshore and 95 megatonne offshore.

In an attempt to maintain exports, China has embarked on a crash development of shallow-water offshore wells. The oilfields off her eastern seaboard contain an aggregate resource estimated to exceed that of the North Sea. Moreover, they offer the advantages of being further away from the threatened northern border, but near coastal regions where dense population, concentrated industries and all export terminals are located. Indeed, much higher priority is being attached to offshore than to onshore exploration. Her expertise in geophysical prospecting has been examined in Section 2 of the previous chapter. Drilling and extraction are the weak links, although she has insisted that, in the joint ventures, preference for her own equipment will be sought. A 4 200-metre well in the East Sea is the deepest drilled to date. Her first drilling rig, from Hongqi Shipyard at Dalian, saw service in 1971. The Shipyard is now building several self-raising drilling platforms. For deep-water jack-ups and large semisubmersible rigs, she has still to turn to outside sources, recent purchases having been from Japan and Norway, France, the US and Singapore. In 1980, one of these imported rigs overturned with the loss of many Chinese lives in the Gulf of Bohai. The tragedy was caused as much by technical inexperience as by bureaucratic negligence. In this Gulf prospecting has essentially been completed, and exploration and appraisal wells are being drilled, mostly by Chinese personnel. A total of 24 wildcats will be sunk and it is hoped that full-scale production can commence by 1985 or 1986.

Research related to offshore structural engineering, concerning eg, seawater corrosion, platform and pipeline construction, sea waves and thermally induced resonance phenomena (p 96) has been accelerated. As an example, a group at Dalian Institute of Technology is studying the dynamic response, under the forcing of waves, of a frame of universal-jointed columns that is hinged at the bottom. The tension-leg structure for use in deep water is also under examination.

Unlike in the offshore case, China has achieved all-round competence in onshore exploratory drilling and field production, although many of the techniques in use would be talked of as being oldfashioned by Western engineers. In 1973 the Institute of Powder Metallurgy in Beijing succeeded in producing her first synthetic diamond drill. Southern Institute of Iron and Steel Research under the Ministry of Metallurgical Industry cooperated with Capital and Anshan Iron and Steel Companies to build, in 1976, 55-kilogram drill-pipe. During the same year her first hydraulically controlled blowout preventer was made by Chongqing Machinery Plant. The Model 130–3 deep-well rig, mass-produced since the late 1970s by Lanzhou General Machinery Factory, is one of her many outstanding designs, being light, compact and efficient; most heavy oil and gas drilling equipment comes from Lanzhou which boasts three petroleum machinery works. A recent innovation is interesting in being the application of an outcome of research at the 'back end' of the petroleum industry to its 'front end'. A structural adhesive, developed by the Harbin Institute of Petrochemistry, was found to be suitable for bonding tungsten carbide alloys to the ends of steel drill bits used at Daqing Field. The national record drilling depth is 7 002 metres. Research departments are retained by all oilfields, an arrangement efficient for the solution of local problems regarding geology and production conditions. At the 1978 All-Nation Science Conference, the Chinese claimed to have their 'own inventions in the science and technology of the oil industry, and in some fields have caught up with or surpassed advanced levels in other countries'. Some of these inventions were described at, for instance, the International Meeting on Petroleum Engineering convened in Beijing in 1982. The technique of secondary recovery through water injection at an early stage in low-pressure deposits, which they have pioneered and refined to a high level, is predominant in field production; their largest truck-mounted injection pumps work at 100 MPa. However, this technique forces the field to peak early and ultimately may limit the amount of total recoverable resources. In many mature oilfields the flows now consist of more water than oil; recently US consultants were brought to Daqing to advise on this problem. Urgent efforts are being put into research on methods of raising recovery efficiency. One project, undertaken at the Department of Mechanics in Beijing University, is to study the rise of a cone of water in porous media towards the production well situated above the oil-water interface. The improvement of oil and gas recovery in general has been specified as a key area of work in the 1978 Science Programme.

Also mentioned in the Programme are ways to make full use of low-calorie fuels. In one particular case, namely full-scale field production of shale oil, Chinese expertise is likely to be already unparalled. At the Fushun Open Pit Coal Mine (section 3 of previous chapter), part of the overburden overlying the coal is oil shale and is processed in retorts on the edge of the pit. The initial setup was credited to the Japanese when they were in occupation, but many innovations have subsequently been made, especially by the research institute that has been established nearby. The Dalian Institute of Petroleum Research is another relevant centre. Besides carrying out routine composition analysis of shale oil by chromatography and infrared spectroscopy, it has studied their catalytic cracking, solved the problem of nitrogen compound poisoning arising therein, and developed the process of equilibrium flash vaporization. At Maoming in Guangdong, six retorting plants produce a better grade of shale oil than Fushun. 'Waste' matter such as spent shale, slag and gases are further utilized in making substances like chemical fertilizers and cement. Recently, a new type of oil shale dry distillation furnace and a technique of preparing oil shale dust for combustion in boilers have been announced.

As in the minerals and coal sectors, transportation is a major stumbling block. There were only about 70 kilometres of oil pipelines per megatonne of extracted crude in 1980. The network connects Beijing and Tianjin (1) through Qinhuangdao with Shenyang and thence separately runs to Daqing, Fushun, Dalian and Korean border, (2) with Shengli oilfield and Nanjing in the southeast, and (3) in the west through Lanzhou to Yumen oilfield and from Urumqi onto oilfields near the border, the connection between Yumeng and Urumuqi being by rail. Most pipelines in this network are 0.62 metre in diameter; some sections have sophisticated heating systems because of the generally high pour point of Chinese crude. Construction has commenced in 1977 on a 1 100-kilometre strategic pipeline linking Lhasa in Xizang with Golmo in Qinghai, near where are several oilfields. The Chinese knowhow in pipelaying must not be dismissed lightly: like the railway being built there in parallel (section 1, next chapter) this newest pipeline will serve as a concrete proof of their ingenuity and inventiveness. The major difficulty for them lies in the inability of their steel industry to supply enough high-quality steel and in the lack of seamless pipe manufacturing capacity. In fact, no seamless tubings exceeding a metre in diameter are yet produced. Natural gas pipelines are few outside Sichuan province, which has a network of some 1 200 kilometres. The situation in offshore development is more satisfactory. Some seafloor pipelines are envisaged in long-term plans. The 'Hungqi' in Dalian as well as other shipyards in Shanghai and Qingdao can easily fabricate tankers of 50–100 kilotonnes capacities, appropriate for short-hual transportation. When petrol and diesel oil were first distributed to Xianggang the transports used were tank railcars, but in the last few years they have exclusively been Chinese oil tankers. Incidentally, it may be noted here that shipbuilding is in the hands of the China State Shipbuilding

Corporation, set up in 1982 to replace the former Sixth Ministry of Machine Building. The Corporation has inherited the thirty or so research and design institutes of the Ministry.

Another weak link, one which has persisted since the 1950s, concerns refining. The country can produce the full range of refined fuels and oils, though most crude is processed into diesel (33 per cent) and petrol (17 per cent), the balance being made up by kerosene, lubricating oils, paraffin wax, white oil and petrochemical products. About 10 per cent is lost per unit quantity of petrol processed in China's plants. Her difficulty lies not so much in technological level than in aggregate capacity. Thanks to the withdrawal of Soviet advisers who took all their blueprints away, she did not begin making refinery equipment by herself until 1964, after obtaining some assistance from Romania. Although the growth rate in refining capacity was consistently high in the two decades by world standard, in absolute terms it has always fallen short of field production volume, the increase of which, admittedly, has been phenomenally fast. The purchase of relatively large refineries was the first of (and in monetary terms accounted for 26 per cent of) whole-plant imports in the early 1970s (rising to 48 per cent in 1978). The largest installation is that at Daqing with a crude throughput of 6 to 8 megatonnes each year, equal to about a quarter of the capacity of the biggest in the West or little more than a tenth of Daqing's annual field production. The rest of the crude output is moved through pipelines to Beijing or Dalian for processing or export.

The difficulty in capacity may now be easing with the stagnation or fall of field output but is likely to reappear when offshore oil starts flowing. A little help is afforded by the continuing expansion of storage facilities, though. The biggest of them, which, of course, carries strategic implication, may be the underground granite cave 0.1 million cubic-metres in volume at an undisclosed location. Incidentally, the country probably has the most active programme of subterranean construction in the world, if those of a strictly military nature are not to be counted. In the northern provinces, there is a long tradition of storing foodstuffs underground. Evacations near Xian of palaces built during the Qin dynasty of the tenth century BC have uncovered wells with clay food containers at their bottoms. In the south, as an example Chongqing, suffering from a hot climate and high humidity, is a city with several refrigerated warehouses that offer an aggregate capacity of 10 kilotonnes. Nationwide, many civil defence shelters have been converted into granaries, factories, schools and even hotels.

The Ministry of Petroleum Industry sponsors an Academy of Petroleum Research, which contains a number of research institutes. Refining processes are studied at the Institute of Crude Oil Refining, Institute for Petrochemical Research, both in Beijing, and Dalian and Lanzhou Institutes of Petroleum Research. (The latter three also undertake petrochemical work; the one at Fushun with a similar name is now more oriented towards research on environmental protection.) Within the Academy of Sciences system, the Heterogeneous Catalysis Laboratory of Dalian Institute of Chemical Physics is distinguished for preparing over a dozen of heterogeneous as well as homogeneous catalysts for synthetic liquid fuel. Some of the homogeneous catalysts have been incorporated into the fluid catalytic cracking process that was initially developed at the Beijing Institute for Petrochemical Research and since then commercialized at Daqing. Incidentally, this chemical physics centre is also looking at the direct conversion of chemical energy into electricity in fuel cells. Founded in 1949 and formerly called the Institute of Petroleum, it got the present name in 1958 when some of its technical personnel were transferred to staff the Lanzhou Institute of Chemical Physics and Shanxi Institute of Coal Chemistry. The Lanzhou Institute has a division engaged in analytical chemistry, and which has succeeded in identifying individual components of light fractioning of oils from many oilfields all over China. As for the Shanxi Institute, the central concern is coal research, as said in the last section. However, since the mid-1960s accompanying the dramatic increase of national oil production, petrochemical work has been pursued there. Advances were achieved in many areas, including the cracking of crude oil in olefine, hydrorefining of cracked gasoline and isolation of isoprene from C-5 fraction. Recently started is the work on fibre-supported catalysts. Incidentally, Chinese crude generally has high wax but low sulphur contents, so that hydrorefining is not critical.

The Chinese possess the capability to design, build and equip petrochemical complexes, the Beijing General being one of the results of such efforts. Although their technological lag remains considerable, their repertoire of home-produced lubricants and petrochemical feedstocks is fairly complete. Their general philosophy is to avoid the waste of hydrocarbons in being simply burnt as fuels and to conserve them as feedstocks for plastics, chemicals, fertilizers and drugs manufacturing processes, though to this day about 40 megatonnes are still burnt annually. The biggest ethylene plant, with a designed output of 0.3 megatonnes per year, was completed in 1981. Significant quantities of lubricating and non-lubricating oils, mineral jelly and wax are now exported.

Research on the petrochemical front is partly in the hands of establishments under the central administration of the Academy of Sciences or the Ministry of Petroleum Industry. Under the Academy, the Dalian Institute of Chemical Physics has prepared catalysts for olefine polymerization and for application in the antifreeze production plant belonging to Dalian Petrochemical Works. A new reforming catalyst was obtained that gives 55 per cent (10 per cent more than before) aromatics from Daqing oil. Work is continuing on the dehydrogenation of long-chained paraffins by the bimetallic catalysts platinum-gold and platinum-silver. The oxidative dehydrogenation of n-butenes to butadiene catalysed by bismuth-molybdenum oxides and bismuth-molybdenum silica is being studied at the Lanzhou Institute of Chemical Physics, which has also developed a number of solid lubricants and special coating materials and devised routes for their syntheses. A current project is to test primary amines as ashless additives to lubricating oils. Petrochemical syntheses via rare-earth catalysts is a strong area at the Institute of Applied Chemistry, Changchun (section 5 of previous chapter). Chengdu Institute of Organic Chemistry concerns itself mainly with the chemical utilization of natural gas. Its six laboratories take the respective titles of Homogeneous Catalysis, Heterogeneous Catalysis, Polymer Chemistry, Fine Organic Synthesis, Analytical Chemistry and Plasma Chemistry. Work completed has covered the manufacture of vinyl chloride via dilute acetylene from natural gas, catalysis of natural gas reforming, production of acetylene from crude through plasma jet, etc.

Under the Ministry there is a Sichuan Institute of Natural Gas Research (mentioned already on p 122 in the preceeding chapter) that conducts, as the name implies, research connected to natural gas and where some projects are in petrochemistry. Another ministrial establishment, the Institute for Petrochemical Research, has contributed in large measures to the topics of platinum/alumina fixed-bed reforming, reforming via bimetallic catalysts like platinum-tin/alumina, and lubricating oil hydrogeneration with cobalt and nickel-molybdenum catalytic agents. In addition, many petrochemical plants maintain research laboratories, where often the only difficiency in evidence is the lack of microprocessor-based analytical instrumentation. However, no space will be devoted to an examination of the work at these establishments and institutions, which is generally speaking outside the scope of 'energy'; the sector of chemical fertilizers has already been covered in Chapter 4.3.

7.3 Electricity and Hydropower

Electricity production in China is patently inadequate. The overall shortfall may be as much as 25 per cent of the demand and this, besides causing hardship to households in many cities and villages, acts as a universal constraint to industrial growth, a specific example being the case of Wuhan Steelworks mentioned in the introduction to this chapter. The outlook is a highly sobering one. Between 1952 and 1980, the national electricity generating capacity grew by annual average of 13 per cent while generated power rose by 14.2 per cent; in 1981 and 1982, the annual averages were 3 per cent and 2.9 per cent respectively. Indeed, the economic plan promulgated at the Fifth National People's Congress disclosed that one of the main reasons for holding down steel production was to divert more power to light industries. Caution will be needed, however, not to enter a vicious circle because steel is required in the construction of power stations and the development of fuel transport systems, factors that determine the expansion of generating capacity and the increase of realized output respectively.

The state department with overall responsibility for the electricity generating industry is the Ministry of Water Conservancy and Power. However, some factories making small boilers and steam and hydraulic turbines are administered by the Fourth Production Bureau of the former First Machine Building Ministry, the Eighth Production Bureau of which controls some plants manufacturing small electric generators and power distribution equipments.

The shortage of electricity arises not only from a scarcity of installed capacity because of underinvestment and an underproduction due to tightening fuel supply, but also from a lag in generation technology that is closing but still significant at present. One contributing factor to this lag is the inadequate supply of high-quality steels. The biggest super-high pressure boilers in the country, products of the Shanghai and the Harbin Boiler Plants, have unit capacities of only around 1 000 tonnes per hour. The largest thermal turbogenerators are at Liancheng; their rating is 400 megawatts, or less than a quarter of that of the world's most advanced ones. The Steam Turbine Factory of Shanghai may have completed the design of 600-megawatt units after overcoming a great many obstacles, and that of 1 000-megawatt sets are being tackled. The state-of-the-art gas turbine generators are rated at 10 megawatts, ten times less than the specification of serially available sets in the West. (Hydro turbogenerators will be discussed later.) There is, nevertheless, an area where the Chinese have the technological edge. They constructed the world's first twelve-megawatt thermal generator with water-cooled inner rotar and stator in 1958, the first such set rated at 300 megawatts in 1972, and over the years have successfully applied the technique of water cooling in all large thermal and hydraulic units.

In the scope of auxilliary power production there is an active programme of research on magnetohydrodynamic generation. One of the ten laboratories constituting the Institute of Electrical Engineering in Beijing is wholly devoted to this investigation. Its magnetohydrodynamics (MHD) device, equipped with super-conducting magnets of the dipole saddle type, has attained electrical output of 595 kilowatts for short durations and 23.6 kilowatts over extended periods. An open cycle MHD-steam combined generator is also under trial. The Institute of Engineering Thermophysics, also in Beijing, is engaged mainly in basic research on aviation turbojets, activities that will be described in Chapter 8.1, but some work is supported there on industrial turbomachinery. A combined cycle gas-steam power generator is being tested there. In addition, a programme on energy conservation techniques including the application of heat pipes has recently been inaugurated. Also noteworthy is a joint project by Shanghai Power Equipment Research Institute and Shanghai Electric Machinery Manufacturing Works on superconducting alternator. The model developed so far has niobium/titanium wire-wound rotor and has been run at 3 000 rpm as a synchronous condenser while connected to the local grid, attaining a peak rating of 428 kilovoltampere. The general structure of this experimental model is similar to that of the 2-megawatt machine built and studied by America's MIT.

China has about 120 fossil stations rated at 30 megawatts or above, which between them provide some 70 per cent of national electricity production. Natural-gas-fired plants, found mostly in Sichuan, are small by Western standard and unable to benefit fully from the economy of scale. There is now a nationwide ban on commissioning large oil-fired boilers, as part of the plan to revive the general dependence on coal (see previous section). In electricity generation, coal already provides 80 per cent of the input. The biggest thermal power station is the Jinghe in Liaoning province. Coal-fired, it has a designed capacity of 1 100 megawatts or a third of the limit in the world. According to the 1978 Science Plan, research is to focus on technical problems in the erection of power plants at coal pit mouths.

Mentioned in the Plan was also the work on problems in building major hydropower stations, relating to geology, hydrology, meteorology, reservoir-induced seismicity and engineering protection. Coal production and transportation bottlenecks are virtually certain to restrain the expansion of coal-based power supply throughout the coming decade, whereas hydroelectricity is a comparatively untapped resource (Table 2). The latter, a renewable, safe, nonpolluting and environmentally often acceptable source of energy, to this day provides only 3 per cent of national primary power production or roughly 18 per cent of electricity output, though the prospect is to raise the 18 per cent to 25 per cent by the end of the century.

The Chinese have constructed about 70 hydro stations of 30 megawatts and over. They used to view hydro development as capital-intensive and having too long a lead time, involving extra work on dams and transmission lines. Nevertheless today, together with the coal sector, it is enjoying top priority. As in coal, much hope is attached to development in the southwest, where hydropower potential is greatest. Indeed, in Guizhou which has a potential put at 18.8 gigawatts, nine plants with an aggregate capacity of 4 500 megawatts are being built on the River Wu, which at present is attended by a sole 630-megawatt station at Wujiangdu (see below). A massive scheme for the Hongshui River, a section of upper Pearl River in Guanxi province, has been under study for many years and seen many ups and downs including bureaucratic struggles between ministries with formal responsibility for different aspects of the project. A small 60-megawatt station at Ertan and the first 400-megawatt stage of the installation at Duhua near completion, but the overall scheme calls for again a total of ten hydroelectric plants along a 1 100-kilometre stretch of the river. Preliminary plans are also being formulated concerning the upper reaches of the Yangtze River. An aerial survey was conducted in 1980–82, leading to the estimate that the upper reaches have 70 per cent of the nation's potential hydroelectricity resources. A feasibility study has been initiated on a huge hydropower project on the Yalong River, the main tributary in the upper reaches of the Yangtze. Massive developments are already under way or in an advanced stage of planning for the middle reaches of the Yangtze in central China (see later).

Again as in the case of coal, small-scale sites are responsible for a significant proportion of the production capacity (equal to 24 per cent in 1982). These small hydropower stations number roughly 100 000 and are often integrated with native irrigation and flood control systems; some are rated as low as 0.4 kilowatts. The programme for their construction, started in the Great Leap Forward years and continued during the Cultural Revolution period after a lull in between, was aimed to encourage self-reliance on the local level. The Chinese are the most experienced of all people in the utilization of mini-hydroturbine/generator units; they recently sold three such sets to Intermediate Technology Industrial Services Ltd of the UK.

At the opposite end of the scale is the Liujiaxia hydropower harnessing project. A product of indigenous engineering ingenuity, the station built has an installed capacity of 1 250 megawatts and is at present the largest on the

Yellow River. The Gezhouba, when completed on the 1986 target date, will be even bigger at 2 715 megawatts. Again entirely Chinese in design and construction, it represents the first major hydroelectric project on the Yangtze, being 3 kilometres upstream of Yichang city on the middle reaches. Its first stage was completed in the summer of 1981 and work is continuing on the second stage. The construction will provide experience for a future mammoth 'Three Gorges' project to erect another station upsteam, the probable capacity of which has been variously put at 15 000 to 25 000 megawatts. Hydraulic turbogenerators made in China are predominantly of the Francis type, although adjustable-blade sets chiefly of the Kaplan design are found depending on the heads encountered. Liujiaxia houses the present largest one, rated at 300 megawatts and having water-cooled inner stator and rotor. Gezhouba has two low-head 170-megawatt and (eventually) nineteen 125-megawatt units. The 170-megawatt sets, each weighing 4 000 tonnes, have been jointly made by the Dongfang Power Equipment Factory and the No 2 Heavy Machinery Plant in Sichuan. The maximum design head is 27 metres, the peak rate is 1130 cubic metres per second and the runner shaft has a diameter of 11.3 metres. The Longyang station (of which more in the next paragraph), also under construction, will have four turbogenerators. They will be rated at 320 megawatts each and made by Dongfang Generator Plant in Sichuan; the first one is scheduled to be in place by 1985. Design has been completed for 500-megawatt units, which are planned for use at the 'Three Gorges' station.

The country is quite advanced in dam construction technology. She has some 86 000 dams, most of which are small and usually of the earthfill type, almost always built by a sluicing-siltation method that was devised by the Chinese. Less common are the stone masonry and rockfill types, the latter often constructed by directional blasting technique. The last named technique has been studied at the Institute of Mechanics under the Academy of Sciences, and its application extended to making farmland by hill levelling. Least common are gravity concrete, hollow gravity and concrete arch types which are met among the larger dams. About 14 000 have heights in excess of 15 metres, representing nearly half of the world's total number of tall dams. The two highest are at Longyang (175 metres) and Wujiangdu (165 metres). The one at Longyang Gorge, on the upper reaches of the Yellow River in Qinghai province, is of the arched concrete type and, when completed, will create the largest in volume (26 500 cubic metres) artificial reservoir in China. The Wujiangdu dam, across the largest river in Guizhou, is of the arch gravity type. It has a complex formwork and ski jump spillways, the design of which was a great achievement. Furthermore, at depths from 80 to 200 metres below its site were karst caves; to ensure stability of the foundation Chinese engineers forced 82 000 cubic metres of cement into these caves under high pressure. Such a method of stabilization has subsequently been adopted in dam construction over limestone areas.

Along the Maotiao River, a tributary of the Wu, there are six medium-size hydro stations, situated on geological formations that have also entailed unusual problems. At the third station downstream, for example, water pouring out of sluice gates has to go over the top of the power house. The double-curve dam of the fourth station was built on top of an arch bridge resting on the cliffs, because of the undesirable presence of a 27-metre layer of sand and gravel in the river bed. The dam at Gezhouba spans 2 561 metres, reaches 47 metres in height and incorporates a new design to prevent silting. As much as 50 megatonnes of sand and mud pass there each year and the operation of shiplocks enhances the likelihood of silt accumulating below the dam. Special retaining dykes have therefore been built to divert silt into positions where it can be flushed away when scouring sluices are opened periodically. The dam structures stood up to a severe test soon after completion, when in July 1981 they safely disgorged flood waters flowing at 72 000 cubic metres per second – the highest recorded flow since 1896. Incidentally, remote sensing has been used by Chinese geologists in dam site selection, a recent example being the case of the 60-megawatt Ertan Station in Guangxi. An illustrative case of the use of satellite imaging is provided by the Datong station in Shenxi.

In contrast, electricity transmission technology is not a strong point. Nearly 15 per cent of energy is lost in generation and transmission in China, which is still rather backward in the design and production of high-VA transformers and circuit breakers. The 470-MVA transformers manufactured by Xian Transformer Plant appear to be the biggest of serially produced units. The 1978 Science Programme has identified super-high-tension power line as a subject of intensified research. The largest existing switchgear at a hydropower installation is that at Liujiaxia, for which all cables (oil-filled and aluminium-wrapped, totalling 550 circuit-kilometres), power transformers, circuit breakers (air-blast), lightning arresters and porcelain insulators were domestic products. The transmission there is at 330 kilovolts; provincial ties typically go to 220 kilovolts only. The 500-kilovolt 750-MVA and 1 500-MVA transformers, mutual inductance devices, protective relays and control instrumentation used in the Yaomeng Power Plant (in Pingdingshan Coal Mine) to Fenghuangshan Transformer Station (adjacent to Wuhan) transmission line have been imported from Sweden, France and Japan. This line connects the provincial grids of Henan and Hebei, and works at 500 kilovolts which is the present limit in the country. Its extension from Fenghuangshan to Gezhouba, passing 1 170 metres across the Yangtze, was completed in 1982 largely through self-efforts. Shenyang Transformer Plant and Cable and Wireless Factory are testing prototype 750-kilovolt systems; by comparison, interstate ties in the US were upgraded to 765 kilovolts in the late 1970s. In the case of direct current transmission, which offers advantages for point-to-point transfer of large blocks of power over long distances, a blank persists in China. Her main difficulty apparently relates to the technology of converters and especially high-current high-voltage thyristors. Regional power grids span only the north, northeast, east, central and the three provinces around the Lanzhou industrial complex. However, a long-term plan for the nationwide integration of provincial grids is being implemented, with the work in the south started first. Notably, the Chinese were the first to practise repairing high-tension wires under live potential.

A number of research institutes exist under the Ministry of Water

Conservancy and Power. Two are at Zhengzhou and Wugong, but their activities focus on purely hydrological matters. Others are more oriented towards work concerning hydropower stations. Among them, the Water Conservancy and Hydroelectric Power Exploration and Design Institute at Changsha has participated in the construction of eg, Longyuan Hydropower Station mentioned above, and the one at Kuming, in that of Lubuge Station. The Ministry's Tianjin Institute of Design has made outstanding contributions to the subject of highhead spillways. A design has evolved by which pitting is prevented through a combination of approaches such as adopting an optimum profile for the structure, providing aeration and utilizing special materials in the construction. From the Institute of Yangtze River Water Conservancy and Hydroelectric Power, a group received a state prize in 1980 for developing a new concrete that has low exothermicity and expansion coefficient. This institute in Wuhan has been heavily involved in the Gezhouba project. The Nanjing Institute of Hydraulic Engineering has a department where interests concern the hydrodynamics of sluices and spillways in hydroelectric installations. Partly funded by the Ministry of Communications, the Institute also supports research into the morphology of navigable waterways, silting of ports, tidal erosion of river estuaries and stability of dyke slopes. A series of publications have emerged, reporting a systematic investigation of laminar and turbulent flows in pipes and open channels. With turbulent flows treated stochastically, generalized formulae for both types of flows have been derived for the velocity distribution, drag coefficient, etc.

The largest of research establishments is the Water Conservancy and Hydroelectric Power Research Institute in Beijing, first mentioned on p 80 Special in being subordinate to both the Ministry and the Academy of Sciences, it was founded in 1958 as a result of the merging of three organizations: the Water Conservancy Research Institute of the then Water Conservancy Ministry, the Hydropower Research Institute of the then Electric Power Industry Ministry, and the Academy's Department of Hydraulic Research. Abolished in 1969, it was re-established in 1978 and now consists of eleven divisions. Their titles are: Water Conservancy, Water Resources, Sedimentation, Hydraulics, Structure and Materials, Rock and Soil Mechanics, Seismatic Engineering, Hydraulics and Electric Machinery, Automation, Cooling Water in Thermal Power Plant, and History of Water Conservancy (work at the last division being chiefly the collection and interpretation of relevant historical data). There are also an instrument factory and a computer centre. Some of the work at the first three divisions concerning hydrological geography and sedimentation fluid mechanics has been sketched on p 101. Examples of recent accomplishments of the other divisions in the field of hydroelectricity are: introduction of water reduction admixture for concrete and polyurethane grouting of hydro structures, elucidation of dynamic behaviours and seismic responses of the power station inserted within the Fenshuba dam, mitigation of cavitation damage to turbine blades, solution of vortex induced vibration and air entrainment problems, calculated of elastic water-hammer effects on governing stabilities, and computation of pressure distribution in tunnel spillway flow with the help of finite elements. Present emphasis is on studies of: high-dam structures and new materials for hydraulic construction, high velocity flow problems arising in spillways and outlets of tall dams, rock and soil mechanics in major projects, and design of large hydro turbogenerators.

Earthquake resistance of hydropower structures constitute an important topic of study as many parts of the country have high seismicity. Earthquake engineering has been discussed, however, previously in Chapter 5.6.

7.4   Nuclear Energy

China is richly endowed with uranium deposits, amounting to 0.8 megatonnes of the metal (Western estimate) against the world's proved reserves of roughly 3 megatonnes. Most are in the southeast and the extreme northwest, although recently a major find has been announced in a volcanic region of the southwestern province of Yunnan. Despite the availability of this resource, it is really debatable whether the government should opt for fission power generation, which has the characteristics of being highly centralized and capital intensive and has uncertainties regarding safety, health hazards from accumulative dose of low-level ionizing radiations as well as radioactive waste disposal. Nevertheless, in 1980 participants at the inaugural meeting of the Nuclear Society of China recommended an accelerated development of this form of energy; shortly later, the State Scientific and Technological Commission proposed that by 2001 nuclear power generation should exceed 10 000 megawatts and suggested the construction of six nuclear power stations by 1991. However, final approval has been given for only two, in March 1983. Of the pressurized water reactor type and rated at 900 megawatts each, these two are to be in Guangdong, within the Shenzen 'Special Economic Zone' from where electricity can be distributed to nearby Xianggang, which is in this context perhaps too near (for safety). Their erection replaces a hydro scheme originally planned in the province.

It remains unclear to what extent China, if and when she does go nuclear with hastened pace, will rely on foreign reactor expertise. Concerning the Shenzen station the purchase of complete reactor-generator assemblies has been discussed with France, West Germany and the US. For political reasons the deal with France seems to have the greatest chance of going ahead. On the other hand, she is fairly experienced in the operation of small reactors (see below) and intimate outside involvement may possibly be avoided in this sector of potentially strategic significance. Since January 1983 she has been erecting a 300-megawatt prototype power plant of her own design at Qinshan, 130 kilometres southwest of Shanghai and facing the Hangzhou Bay. Scheduled for completion in 1990, the plant will be protected by a dam 8 metres high and 1 700 metres long, furnished with a Chinese pressurized water reactor, and built to specifications giving a greater 'safety' margin than at similar installations abroad. It may be relevant to note that, from the early 1970s onwards, she has also been developing submarines that are powered by nuclear reactors. Designated the 'Han' class, the first boat entered service in April 1981.

China is even more advanced in uranium extraction and processing technology. Work in this field was originally controlled by the Second Ministry of Machine Building responsible for thermonuclear weapons development, which has now been replaced by the Ministry of Nuclear Energy that centrally coordinates both such work and activities in the reactor power station sector. Not much is known about the Uranium Geology Bureau in the Ministry. It was reported, however, that airborne exploration has been conducted, in combination with aeromagnetic and aerogravity surveys (section 2, previous chapter), on a scale of 1 : 200 000 or less for most parts of the country and followed, no doubt, by detailed field prospecting in selected areas. The country now operates three principal uranium mines. At the 1981 International Conference on Solid State Nuclear Detectors, a Chinese paper described a prospecting method by the use of track etch technique to monitor surface radon. The authors disclosed that by this means deposits were located at a depth of 200 metres or more, when conventional surface radiometric techniques had been ineffective.

The country's first ore concentration plant was set up in 1957 at Zhuzhou in Hunan province under some technical assistance from Czechoslovakia. In 1963 the Lanzhou Gaseous Diffusion Station began processing uranium-235, using power furnished by an adjacent hydroelectric facility. Westerners in 1980 visited, openly for the first ime, a plant which refines 1 000 tonnes of concentrates annually to 'nuclear purity' uranium dioxide that may, when necessary, be enriched at another plant to produce weapon grade material. Gaseous diffusion is still the main method favoured by the Chinese, but a gas centrifuge plant has been planned or may even have been built, and new techniques such as laser isotope separation are being studied. Many research activities are concentrated in Lanzhou but most of them are classified.

The Ministry of Nuclear Energy runs at least five research establishments, named simply First, Second, ... and Fifth Research Institutes. They are described as working on scientific and engineering problems connected with the exploitation of nuclear energy and with the applications of radiation and particle flux in materials technology, agriculture and medicine. No detailed information is available, however.

Since 1958 China has operated experimental reactors for basic research. The oldest one, bought originally under the Sino-Russian 1955 Agreement, was acquired in 1958, redesigned in 1962, upgraded from 6.5 to 10 megawatts in 1976 and recently reconstructed again. Sitting at the Institute of Atomic Energy, this reactor uses 2 per cent enriched fuel rods and heavy water as moderator and coolant. The Institute is located in a southwestern outskirt of Beijing, away from universities and other research centres most of which are found in southwestern suburbs. Founded in 1950, it has as its predecessor the Institute of Modern Physics, but was renamed and put under the dual direction of the then Second Machine Building Ministry and the Academy of Sciences. More is known about it because of its link to the Academy. It now comprises over twenty laboratories devoted to the broad areas of nuclear physics, radiochemistry, as well as reactor physics and technology. Current programmes in the technology area include investigating thermal hydraulics, radiation damage in steel and general material problems in a large power reactor that is simulated by the 10-megawatt reactor. Since 1972 the Institute has also served as the chief producer of over 90 radioisotopes for nationwide distribution. The two separators used in the production are of the old-fashioned electromagnetic type, but have homemade ion sources and collectors, one of the latter being a PIG type discharge device with graphite slit for extracting ions. The Institute houses a second reactor which is an entirely indigenous installation. A 3.5 megawatt swimming pool type containing 3 per cent uranium 235 surrounded by beryllium and carbon reflectors in aluminium rods, it went critical in 1972 and is involved in tests on advanced fuel cycles, spent fuel reprocessing and pressure vessel steel. Waste is disposed of by way of straight dumping into the river – after treatment has reduced the radioactivity to 0.01 microcuries per cubic metre.

The Institute of Nuclear Research, set up in 1964 in the northwestern suburbs of Shanghai, has been selected for the job of designing the prototype power plant at Qinshan. It has nine divisions engaged mostly in basic research, often in collaboration with Fudan University's Department of Atomic Energy (the only academic institution with such a designation in the country). The Nuclear Chemistry division is a producer of radioisotopes such as potassium-42, gallium-67, arsenic-74, strontium-85, indium-111, etc, as well as labelled compounds and radiopharmacopoeia. In the division for reactor engineering, work includes material corrosion experiments for reactor design evaluation and tests on treatment of low-level radioactive waste liquid with packed-bed electrical dialysis. Some information about the neutron and gamma ray sources at the Institute has already been given on p 66; there is available also a Mossbauer installation, equipped with two 4 096-channel analysers, one of which is indigenous.

China's first high-flux reactor was recently put into operation at Southwest Institute of Physics lately established at Dongshan in Sichuan province. Designed to give 125 megawatts and a maximum flux of 6.2×10^{12} neutrons per square millimetre per second, the reactor has become the second largest producer of radioisotopes in the country. Fusion is also studied at the Institute.

The Chinese are optimistic about fusion energy, which is safer and cleaner but which presents formidable scientific and engineering difficulties. A medium-size Tokamak has just taken shape at Southwest Institute of Physics. Earlier, a 2-tesla Tokamak (the 'CT-6B') was built in 1973 and completed in the following year at the Institute of Physics in Beijing. Over 2 500 discharges have been monitored to date; the highest plasma temperature reached is 3.5 million K. Since 1978 a microprocessor-based data acquisition and processing system has been attached to the CT-6B. A 100-kilojoule linear theta pinch was built also during the Cultural Revolution years. Plasma heat losses by laser scattering, streak photography with a rotating mirror camera and many other results were obtained with it. Another group there has finished rigging up a 1.5-tesla 2.7-megajoule belt pinch device (the 'GBH-1') with rectangular cross-section and work is proceeding on high beta plasma diagnosis. Furthermore, a new apparatus of the magnetic mirror configuration is being assembled. Already commissioned are the two D-shaped mirror magnets which incorporate multifilamentary niobium/titanium composite conductors, with a copper to superconductor ratio of 2:1 and operated in partial adiabatic stabilization mode. The field strength at the centre of the system corresponding to minimum intensity is 2.5 tesla. The liquefier and refrigerator assembly is of the turbine type with a rating of 100 litres per hour.

Finally, since 1974 laser initiation has been studied at the Institute: LiD-LiT is the current target substance but the compression/heating pulse is as yet able to go up to only 36 joules in 6 nanoseconds. Research on laser implosion had commenced earlier at Shanghai Institute of Optics and Fine Mechanics. Founded in 1964, this place has been the country's foremost centre for research requiring high-power laser facility as exemplified by its 10-gigawatt carbon dioxide pulsed laser; such work is often classified. A recent piece of research reported in the open literature relates to an investigation of the effect of target atomic numbers on the self-focusing and filamentation of the laser beam in the coronal plasma produced. In the inertial confinement programme, neutron release was attained for the first time in 1973 and target compression in 1977. The source employed at present is a six-beam neodymium-glass laser which provides 6×10 joules in 0.1-nanoseconds pulses equivalent to a focused power density in excess of 10 terawatts per square millimetre. Six aspheric lenses and a continuous-wave yttrium aluminium garnet laser are used to focus the beams to within 10 microns. Diagnostics on the laser plasma include X-ray angle-resolved spectroscopy in the 1 to 88 kiloelectronvolts range, electron, ion and neutron spectroscopies, high-speed microphotography, plasma density profile via microscopic holographic shearing interferometry, as well as energy balance determination through laser and plasma calorimetries, laser photometry and X-ray thermoluminescent dosimetry. Two types of targets are currently being tried on, one of which is CD_2 solid microspheres. The other is LiD in glass microballoons of 55 to 85 microns diameter and 0.55 to 0.85 microns wall thickness, filled with either neon at 3 to 9 atmospheres or deuterium at about 10 atmospheres. Incidentally, microballoons from the same source are believed to be involved in research on the sensitization of chemical explosives elsewhere.

The entire Institute of Plasma Physics at Hefei is geared to controlled thermonuclear work. Created originally as a unit within the Anhui Institute of Optics and Fine Mechanics (Chapter 4.5) in 1974, it gained independence in 1978 and is at present divided into a physics section and an engineering section. The former is engaged in theoretical study and numerical simulation of plasma processes, experimental investigation of plasma-wall and plasma-wave-particle beam interactions, measurements of plasma radiation, improvement of plasma diagnostic techniques and high-intensity ion sources, as well as exploration for new methods for plasma heating. The engineering section has research projects on the production and characteristics of ultrahigh, low-temperature and electrically induced vacuums, application of high-current superconducting magnets to fusion reactors, design of high power network pulse supply and energy storage, as well as creation of relevant computer software libraries.

7.5   Unconventional Sources

From foregoing discussion we see that energy productions from fossil fuels and nuclear fission are far from saturation point in China. In one respect, however, this weakness may be turned into a good point, namely the opportunity of building up a power infrastructure based to a large extent on 'soft' and renewable resources. The 1978 Science Programme did include biomass utilization in rural areas, tidal and wind powers, as well as geothermal and solar energies in the list of important topics under 'energy'.

In biogas utilization China was the pioneer and is still now the pacesetter of all nations. Luo Guorui, a native of the southern province of Guangdong, constructed a hydraulic methane tank in 1920 and wrote a book about it: he was probably the first in the world to promote its use. A methane tank built in 1937 in Hebei is still in use today. Its exploitation, sometimes referred to in the Chinese literature as marsh gas production, was encouraged in the Great Leap Forward but not until the late 1970s did it truly mushroom, first in Sichuan, then in all except the coldest parts of the country. A special government unit was set up in 1979 to carry on the promotion among the villages. Today, there are more than seven million methane generating tanks yielding each year 2 000 million cubic metres of the gas, equivalent to 4 500 m kilograms of coal of energy. Most of the tanks are owned by family units and separately connected to biogas stoves and lamps in individual households. Each of these is 8–12 cubic metres in volume and can daily produce 1–3 cubic metres of gas, enough for cooking and lighting purposes for a family of four. Simply constructed, from materials easily available in the countryside, its cost is about a quarter of the average annual rural income per capita, and the plan envisages that by 1990 one will be owned by every family in ten, the fraction now being one family in thirty. (For comparison, in India there are some 0.4 million biomass digesters, the typical unit cost of which is four times the mean annual income, so that by and large only landlords and rich peasants gain from the technology.) With the cosponsorship of several UN agencies, namely ESCAP, FAO and UNDP, the government has set up at Chengdu the Asian-Pacific Regional Biogas Research and Training Centre.

The technology consists of allowing a mixture of animal and human excreta, plant mass and waste water to decompose anaerobically in an airtight container. Additional biodegradable mass is loaded via an intake channel from time to time, but the mixture should be kept near the optimum carbon-to-nitrogen ratio of $1:25$ and (by occasional addition of lime solution or grass ash) a pH value of 7 to 8. The product evolved is 55–70 per cent carbon dioxide, the balance being mainly made up by nitrogen and hydrogen sulphide with certain other gases in trace quantities. Fermentation residues can be pumped out to serve as excellent organic fertilizers (Chapter 4.3). Microbiologists at Beijing Normal University have recently isolated the first pure culture of Methanococci, the bacterium responsible for the methanogenesis. Among the 7 million plus tanks about 40 000 are of medium size with volumes ranging from 50 to 1 000 cubic metres. They are used to power grain and fodder mills, dry farm produce, heat greenhouses or make chemicals such as dichloromethane and tetra-chloromethane. Lastly, about 2 000 are large-sized. Most are operated at commune or production brigade level, and their outputs usually piped to boilers in electric generating stations. The one at Junqiao of Foshan city in Guangdong, for example, has 32 rectangular fermenting chambers to provide daily 1 900 to 2 400 kilowatthours; conditions such as uniform human excreta charge, constant stirring with water pumps and refluxing of fermentation sediments are maintained in order to raise the generation efficiency. A few are installed in towns and cities as environmental facilities, for the purpose of converting organic waste from industry into harmless matter.

Tidal energy potential in China seems relatively insignificant, depending as it does on favourable geological features on the coast. One particularly promising site is the mouth of Qiantang River just beyond Hangzhou. In the Shunde county of Guangdong, a 5-megawatt station on the estuary of the Pearl River exploits mixed tidal and hydropower.

Ocean wave energy harnessing offers greater possibilities, but research is still more scarce in this area. Guangzhou Institute of Energy Conversion (of which more later) has a department devoted to the mastering of ocean energy, where work has only just started. More advances have been made in the direct exploitation of wind energy, though at the moment aerogeneration of electricity is restricted to Inner Mongolia, a region swept by wind coming across the Siberian Plain, and to Zhejiang (and soon Fujian) provinces facing the vast span of the Pacific. According to a study by the National Meteorological Research Institute, over the whole nation an amount greater than the total installed electricity generating capacity can in principle be recovered from the wind. To realize the prospect has become a national goal, and research in this direction is being funded by the State Commission of Science and Technology, Ministry of Electric Power, and Ministry of Machine Building Industry through its Farm Machinery Bureau. On the whole, the main thrust is to develop small-scale windmills for the benefit of local communities scattered over a large area. In Inner Mongolia, since 1977 a pilot scheme has been run involving hundreds of mobile wind-driven generators carried by nomadic herdsmen. Also tested are a few stationary models, designed by the Inner Mongolian Research Institute for Agricultural and Animal Husbandry Mechanization and with a rating of 'several scores of kilowatts'. On Shengsi Island near Shanghai, administratively in Zhejiang province, an experimental turbine with a 21-metre rotor blade and producing 40 kilowatts on average has been under trial for several years. On Pingtan, another island in nearby Fujian, a 55-kilowatt turbine is near completion. All these are small, compared to for example the 3 000-kilowatt aerogenerator on the Orkney Islands, Scotland and the 200-kilowatt turbine at Carmarthen Bay, Wales, but they are probably what will suit China best.

The nation's geothermal resources awaits quantitative assessment but are undoubtedly huge, especially in Xizang where the pushing of the Indian Plate under the Asian Plate results in intensive geothermicity. At the Institute of Geology belonging to the State Seismological Bureau (Section 5.5), under the Deep-Seated Structure and Materials division a geothermics group is exploring fast techniques to measure specific heats, conductivities and diffusivities of underlying earth structures. The main goal is again the development of small electric generating plant. The largest geothermoelectric station, and the first to use untreated wet steam directly from the ground to turn a low-pressure turbine, is situated 89 kilometres northwest of Lhasa in Xizang. The temperature of the steam field there reaches 250 degrees C; the basin is filled with Pleistocene till and glaciofluvial sediments as much as 500 metres thick. The station was conceived after the 1973–6 series of Qinghai-Xizang surveys, completed in 1978, rated at a mere megawatt, but reported to have incorporated a number of Chinese technical innovations. A short while ago a second, 0.3-megawatt station at an adjacent location was put into full operation. Both stations have been connected to Lhasa by a 110-kilovolt transmission line since 1981. A Sino-French survey conducted in 1980 has further revealed three low resistivity layers, presumably geothermal reservoirs lying underneath the general area. The very first prototype geothermoelectric station, however, dated from 1971 and was set up in the less-remote province of Guangdong, at Dengwu of the Fengshun county. Water at 91 degrees C flowing up from an 800-metre deep shaft is degassed and then flash vaporized in two low-pressure expansion chambers, before passing into a 1 500-rpm 86-kilowatt turbogenerator. At Dengwu subsequently another station rated at 180 kilowatts and utilizing binary cycle was built in 1978. The second one in the country to come on steam, in 1974, is on the outskirts of Beijing. Designed by a group including staff at Beijing University, it operates a binary-cycle process of heat transfer with iso-butane as the power fluid. Altogether over two thousand suitable sites have been discovered so far, and in Guangdong, Yunnan, Fujian, Jiangxi, Shandong, Hebei and Liaoning some of these sites have been or are being developed.

As for the direct exploitation of underground water by industry and for domestic heating purposes, the most advanced place is Tianjin, underneath which city is a shallow body of water at 30–53 degrees C and a deep reservoir at 60–96 degrees C, in volume 12 000 million and 7 200 million cubic metres respectively. Further development will be in cooperation with the Danish Board of District Heating, the bilateral agreement having been signed not long ago. At Fuzhou in Fujian to the south, China's first refrigeration plant utilizing subterranean hot water in an ammonia cycle has come on stream in 1982, and is providing 6 tonnes of ice per day. The hot water field there is 5 kilometres by 1 kilometre, at a depth of 200 metres and hotter than 90 degrees C.

Solar radiation incident on the ground in the country has been estimated to average ten thousand million gigawatthours or roughly 1 per cent of that on the entire planet, an amount also equal to the nation's total energy reserves in solid, liquid and gaseous fossil fuels and in fissionable minerals. In 1979, solar cells were installed in a lighthouse of Shanghai harbour to provide power for the beacon. Although large terrestrial installations for the direct energy conversion of sunlight based on high technology have yet to appear, solar cells have been deployed in space since 1971 when the second Chinese satellite was launched. The devices were designed and fabricated by the Institute of Semiconductors under the Academy of Sciences. Work on photovoltaic generation has now been shifted to the Physical Chemistry Research Centre in the Institute of Applied Chemistry at Changchun; cells that have been 'proved' there are of cadmium sulphide or silicon. In the area of heliothermal energy conversion, the Lanzhou Institute of Chemical Physics (section 2) has a research programme on solar desalination. An experimental desalination plant with an evaporation area of 385 square metres is functioning on Wuzhizhou ('Five Finger Islet') around Hainan Island.

Moreover, in the intermediate technology approach China has already scored notable success. This approach first received attention during the Great Leap Forward. For such work Solar Energy Research Institutes have been established by Beijing and Shanghai municipal authorities. Another research and development centre is being built near Lanzhou. Till now, portable solar cookers and, to a lesser extent, solar water heaters represent the most visible impact of alternative technology for solar energy. The year 1974 saw the beginning of mass production of small solar stoves, which incorporate parabolic collectors and have become quite popular in remote areas. A recent innovation is the substitution of plane glass mosaic by a plastic film in the construction of the reflector surface, leading to great reductions in weight and cost. A group at the Academy's Institute of Electrical Engineering is studying the focusing properties of differently arranged sets of planar mirrors mounted along the axis of a parabolic cylindrical reflector. There are now over 30 factories manufacturing such stoves, giving an annual output of nearly 0.1 million square metres in terms of aggregate collector area. The regional Ningxia Institute of Technological Physics is being assisted by Shanghai Solar Energy Research Institute to develop a special model suitable for deployment in pastoral lands of the autonomous region. The Qinghai-Xizang Plateau is another place where the prospect for solar energy exploitation is bright: most locations there enjoy low air density, humidity and turbidity, and receive more than 3 000 hours, or equivalently 8 gigajoules per square metre, of sunshine a year. Back in 1975 the Xizang Industrial Architectural Institute, a research organization under the jurisdiction of Xizang Autonomous Regional Authority, had already introduced a solar oven. Later, the Research Institute of Solar Energy, run by the same Authority, made a portable boiler that could convert 5 kilograms of water into steam within 30 minutes. Recently also a communal solar heated bathhouse opened to service in Lhasa. Besides in this provincial capital of Xizang, many solar water heaters are found elsewhere like in Tianjin and Beijing. A few suburban Beijing villages have integrated communal solar water heaters. An illustrative piece of research in this area is the development, at the Thermophysics Department of the University of Science and Technology of China, of the 'once through' type of heaters which performs better than the natural convection type but is cheaper than the forced circulation kind. In rural districts, in addition to solar stoves 'solar germinating boxes' are also becoming commonplace. These boxes provide concentrated sunlight environment in which crop seedlings will grow faster. The latest new-comer to the repository is the solar energy pump. With a designed capacity of 100 watts, it is the brainchild of Beijing Solar Energy Research Institute and its field test concluded successfully at the beginning of 1983.

The popularization of biogas generators, small-scale windmills and solar stoves is important because, as decentralized energy sources, they serve not only to improve the power supply situation but also to preserve ecological equilibrium in the countryside. An additional desirable effect is the diffusion of basic scientific knowledge and technical skills. At present, two-thirds of rural energy production come from the burning of stalks and straws, at efficiency as low as 10 per cent, which usage accounts for 300 megatonnes or 75 per cent of the annual crop. The stalks should more rationally be returned to the land as natural fertilizers. Another one-sixth comes from the combustion of timber, amounting to 70 million cubic metres or 30 per cent of annual output. Besides losing the availability of timber for other purposes, this practice has at some places resulted in excessive felling of trees that leads in turn to serious soil erosion. Energy development in villages will, as envisaged in a government plan, reduce the burning of stalks by 100 megatonnes and that of firewood by 20 million cubic metres each year in 1990.

The Junqiao biogas installation, the two Dengwa geothermal stations and the Wuzhizhou desalination plant mentioned above have all been designed by the Guangzhou Institute of Energy Conversion. Originally under provincial jurisdiction but transferred during 1978 to the Academy of Sciences, the Institute comprises four divisions working respectively on biomass, geothermal, solar and tidal energy respectively. The Biomass division has research projects on bioconversion for methane, anaerobic digestion, thermal gasification and energy crops. Efforts of the Geothermal division focus on ways of raising thermodynamic efficiencies in phase-change processes. The Solar division is looking at solar heating, cooling and drying, solar boilers, and flatplate concentrators for lower capital cost. The Institute has assisted in setting up several biogas/solar energy self-sufficient villages as well as slaughterhouses and sugar mills utilizing their wastes as raw materials for biogas. There are in addition five small research groups in the Institute. The Windmills group is involved with mini seashore aerogenerators, the Waste Incineration group with gasification on fluidized beds, the Waste Heat Recovery group with operational cost reduction through recombustion and diesel engine supercharging, the Heat Storage group with means of large-capacity storage, and the Heat Pumps group with exploitation of thermal resources with small temperature differentials.

The State Energy Commission, now reorganized as part of the State Economic Commission, established in January 1981 an Institute of Energy Resources at Chengdu in Sichuan. The entire personnel of the Energy Research Department in the Commission for Integrated Survey of Natural Resources under the Academy of Sciences was transferred to staff the new

Institute. Occupied mainly with studying the principles behind the rational exploitation of energy resources (in both urban and rural contexts) as well as technological policies, the Institute has four divisions devoted to: Energy Balance, Energy Utilization, New Energy Resources, and Rural Energy Resources. Three more are expected to be set up soon in: Energy System Analysis, Energy Laws and Regulations, and World Energy.

8 Transportation and Information Transformation

This chapter takes us to the other infrastructural sectors besides energy, namely railway transport, aerospace, information communication and data processing. The ancient Chinese were able artisans and pioneers in astronautics. A passage dated as of the fourth century BC describes a bird constructed from bamboo and wood, which for three days stayed up in the air; kites that carried people were used for signalling during battles before sixth century AD. A relief preserved from the first century AD has been interpreted to be suggesting the idea for the first heavier-than-air flying machine, in the form of a carriage fitted with revolving blades of the screw type and driven by belts – a toy having such blades mounted vertically appeared in Europe two centuries ago and was called the 'Chinese top'. China also invented gunpowder, as was known even in the Middle Ages. The mixture was made in proportions closely resembling the modern standard formula as early as in the second century AD, though not exploited as a solid propellant for missiles until nearly a millennium later. The abacus, in a sense the first (mechanical) computer in the world, existed in the country probably since the same century and definitely not later than the sixth century. Today, however, in almost all of these areas of expertise the Chinese need to catch up with the world's advanced nations before surpassing them once more.

Before looking at aeronautics, we shall below first examine ground transportation. The transport situation is vital for its effects not only on the energy sector but also on the nationwide supply of agricultural produce, merchandise, and raw materials required by industries. Engineering expertise is lacking in connection with highway traffic and aviation but these two remain of secondary importance as far as the movement of civilian things is concerned. Indeed, the annual production of road vehicles in China has actually declined successively in the last two years. Lorries and buses are mainly regarded as constituting merely a supplement to rail and waterborne transports, although technology transfer from the West is sought after, one of the express aims being to improve energy conversion efficiencies of internal combustion engines so that unit consumption of petroleum may be lowered. A joint contract was signed with Volkswagen in March 1983, to produce by 1988 up to 20 000 Santana saloons and 100 000 engines; some of these engines are destined for the domestic market.

Waterway transport, an area where high technology is not crucial (and so like road transport will not be examined in this chapter), works fairly well as it is traditional in China. Her inland waterways total 430 000 kilometres, and of them nearly a third, or over twice the length of her rail network, at present is navigable. The most important of them is the Yangtze and tributaries system, followed by the Xijiang shipping line in Gangdong and Gangxi provinces. The Yangtze itself is navigable for almost 3 000 kilometres: ocean-going vessels can proceed upstream until the Nanjing Bridge (which we shall meet again a few pages on). However, inadequate cooperation among administrative authorities on different stretches of a waterway has given rise to an inordinate amount of transhipping resulting in inefficiency, although this organizational problem is being tackled. Furthermore, most major rivers flow from west to east and, while providing arterial routes between the industrialized east and the interior, act as barriers for road and rail links running north and south. The sole north–south waterway of significance is the Grand Canal, extending 1 794 kilometres and first built 2 400 years ago to connect Beijing with Hangzhou vicinities. Large-scale dredging and other work in its southern section is part of a current programme to raise the shipping capacity of existing routes and to divert water from the Yangtze River to drought-prone areas in the north (p 79).

Highway and water transports are under the adminstration of the Ministry of Communication, once also in charge of the railways, which are now the responsibilities of a separate government ministry. Rail transport is favoured over that on roads because of its greater efficiency, and its ability to use coal or electricity from hydropower in addition to oil, which is needed for export. (It is really a waste to divert oil from petrochemical to diesel and petrol production. The pollution problems of cars and trucks may also soon become important factors of consideration.) Around 80 per cent of freight was carried on trains in the late 1970s; the share has now dropped to 70 per cent but is expected to return to the previous percentage in a few years. The rail sector forms the subject of discussion in the first section below.

8.1 Railways

The impression has been given that the modern sector of the transport system is woefully inadequate in general and, as discussed in Sections 1 and 2 of the last chapter, for the need of the energy industry in particular. This opinion will not be contradicted here. The situation is almost desperate in the case of railways, which carry the burden of inland traffic, to the tune of more than 10 million tonne-kilometres a year. Technical ability is less to be blamed than financial resources allocated for track laying and rolling stock production and maintenance. China ranks second (after the USSR) in the world concerning freight and passenger transport density; her utilization of rolling stock is also rated high. In absolute terms, since 1949 rail freight volume has risen 30-fold, reaching 571 200 million tonne-kilometres in 1981, while track length has expanded by a mere factor of 2.3. In some key sections of the network the capacity today is only 50 per cent of the demand.

Long-haul passenger transport reached 138 300 million passenger-kilo-metres in 1981. On any train on the (Bei)Jin-Guang(zhou) line, which is the busiest and most important route running north and south, one seldom ever finds a 'hard-seat' carriage where, when night falls, but a few people have to remain standing while the more lucky fellow-travellers pack themselves onto seats or cuddle on the floor. One may, on occasions, have the compensation of seeing people who in the spirit of solidarity sleep by rota voluntarily (personal experience).

The growth of the rail network was modest in the 1960s whereas, for political and strategic reasons, efforts during the 1970s have mostly gone into building links with the mountainous southwest – and also into aiding other third world countries like Pakistan, the 2 000-route kilometre Tanzam Railway in eastern Africa that was formally handed over in 1976 being the most conspicuous monument to such self-sacrifices. Of those southwest lines, the 2 200-kilometre track from Xining to Lhasa is yet to be completed, but the 820-kilometre section between Xining and Golmud was opened to traffic in mid-1982, whence Xizang lost the status of being the last province in the whole country to remain inaccessible by rail. The national network now spans over 60 000 route-kilometres, half of which is administered by the Ministry of Railways and half by provincial authorities. Mainline tracks and some provincial lines are of 1435-mm standard gauge. The Ministry has been established with the formation of the government in 1949, but was absorbed into the Ministry of Communications in 1969, only to re-emerge in 1975. Under its auspices are twenty railway bureaux and sixteen sub-bureaux, thirty-three major locomotive and rolling stock factories, as well as the Academy of Railway Sciences. Initial laying of all tracks irrespective of eventual management is in the hands of its engineers, helped by the Army Railway Corps which play a prominent role in many civilian projects like that for the Xining-Lhasa line. However, factories making narrow-gauge rail cars and the laying of branch lines for mines and industrial use are controlled by the Fifth Production Bureau of the former First Machine Building Ministry.

Present orientation of development will bring attention to boosting capacity, uncorking bottlenecks and providing connections with towns and ports that lie close to existing trunk arteries. Containerization is proceeding rapidly: about 100 000 rail containers are already in use and 100 stations equipped with container handling facilities. Heavily-loaded routes are to undergo double-tracking, electrification or both; the Jin-Guan line is to be relieved by additional north–south links formed by the elimination of certain gaps in the network. A plan envisages that in 1983–85 1 700 kilometres of new railways will be built, 1 500 kilometre double-tracked and another 2 000 kilometres electrified, whereas from 1985 onwards till the turn of the century electrification and double-tracking will expand by 13 000 kilometres each, while new line construction proceeds at the rate of 2 000 kilometres each year or double that for the past three decades on average. These policies were spelt out in length at the Symposium on Railway Modernization organized by the Academy of Railway Sciences in August 1980. The Academy has sixteen research departments and five laboratories in addition to an experimental track and test centre east of Beijing. The locodrome is circular in shape but crossed by a straight-line track, enabling motion to be tested at radius of curvature of 200, 350, 600, 800 or 1 000 metres.

The Ministry of Railways is known to run, besides the Academy, at least one college that admits students specializing in railway science and engineering. This college is located at Changsha; a young member of its staff was awarded the 1978 Davidson Prize by the Statistical Laboratory of Cambridge University for having solved, in a 1974 paper, the criterion for uniqueness in a 'q process'. (The work is important to the solution of 'birth and death' problems which may well be relevant to railway scheduling practice. Two authors from the Beijing Normal University have just published an extension of the 1974 result.) A recent contribution in engineering is the design of laser aligned hydraulic rail lifting and lining machine, China's first, which can handle at one time a rail section with sleepers up to 60 metres long.

Civil engineering is one aspect of railway construction in which the Chinese excel; in the previous section on hydropower, we have seen that dam building is similarily a strong point. The 1 100-kilometre Chengdu-Kunming line, with 427 tunnels and 991 bridges totalling 447 kilometres, transverses a region of high mountains, deep gorges, swift rivers and generally difficult geographical or geological conditions. Along many segments of its length landslides can pose a danger, and Chengdu Institute of Geography (Chapter 5.3) has completed a study and their control. The building of the 1 100-kilometre Changsha-Guiyang line has incorporated technical innovations such as drilling hollow bridge piers to hold prefabricated concrete cages into which concrete is subsequently poured, and adopting anchored piles to contain sliding slopes. On the way to Lhasa, China's engineers have pioneered construction across a salt lake. More inventions are expected as the line climbs steeply out of Golmud, to run for most of the remaining 1 400 kilometres at between 4 000 and 5 000 metres elevation, over ground that is in most places permanently frozen to a depth of several metres. Coping with earthquake hazards may already be classed as routine.

Building railways through desert terrain is another area in which the Chinese possess rich experience. The Academy of Railway Sciences has a special desert research department, which has evolved both biological and engineering methods of sand control. They include (1) barrier erection using fences, shallow embankments, ditches and vegetation or tree belts; (2) sand stabilization by pebble, clay or slag coverings and a spray of emulsified asphalt; and (3) guidance by dykes or shields, channels to divert sands with a flow of water, which may then be used to irrigate lineside plants, and a special profile for the formation that prevents sands from drifting against the alignment. The Shapotu Station of Lanzhou Institute of Desert (pp 85–6) has devised a method of supplementing barrier erection. Straw grids are laid on the ground so that the surface wind velocity is lowered by up to 17 per cent. Shrubs and trees planted within the grids are protected by the straw until they take root.

Most mainline tracks are laid to high standard: the philosophy is to minimize track maintenance at source. The ratio of concrete sleepers to wooden ones is about 4:6 in terms of total route-kilometres. The former are reinforced with prestressed steel wires or high tensile strength steel rods, and have a designed failure rate of 10 per cent – 50 per cent when placed under bolted joints – after 50 years in service. A novel type that is shallow but broad in shape has been laid on some 100 kilometres of line. This arrangement provides an almost continuous support without the complications of a continuous slab as well as the retard attribution of the ballast. Sleepers are normally spaced 1 760 per kilometre. Trunk lines are laid with 50-kilograms/metre rail, although the usual grade for less busy routes is 43 kilograms per metre. Since 1978, 60-kilograms/metre rail has made appearance on sections along the lines from Beijing to Shanghai and to Guangzhou. For two decades continuous welding has been the standard practice: strings now up to 1 kilometre long are separated by a pair of 12.5-metre rails with bolted joints acting as a buffer. Until recently, however, technical difficulties were encountered when such rail was laid on long bridges. The problems arose from differential expansions of track and bridge structures and from bending under the weight of a passing train, but were solved by engineers at the track research department of the Academy of Railway Sciences. The Nanjing Yangtze Bridge, an impressive arched bridge carrying a road and a railway, was a proving ground for the engineers. The bridge was at first built with Russian technical assistance but, after 1960, the Chinese began the construction again relying entirely on their own efforts, and finished in 1969. On the other hand, difficulties with points and crossings persist. Movable frogs, the use of which is found to prolong the life of pointwork, are increasingly adopted wherever the angle of the frog is small enough for the turnout to be negotiated at high speed. At a sharp turnout, solid frogs cast in high manganese steel are nearly universally employed.

The average rate of traffic growth is thirteen times that of track laying over the last three decades. This the Chinese has accomplished by increasing the length of trains and the size of freight cars besides improving the power of locomotives. By now a typical freight train is 4 000 to 5 000 tonne; the mean wagon capacity has reached 50.4 tonnes and the figure is still being raised. Trains run at 50 kilometres per hour on average, but the aim is to make 100 kilometres per hour the maximum speed for mineral transportation, 120 kilometres per hour in the case of merchandise and 160 kilometres per hour for passenger service. All these changes will combine to place heavier burden on track maintenance. Mechanization is, therefore, actively sought in maintenance procedures and diagnosis. Machines have been designed and built for tramping, levelling and lining the track, a track recording car has been in service for some years, but technology transfer from abroad in this field is probable.

Electrification is also being accelerated. More than 1 500 kilometres of line have been electrified in southwestern provinces, along routes that encounter steep gradients. Further conversion is proceeding there as well in Shanxi and Hebei provinces, apart from the work on the Guangzhou to Xianggang Railway the first phase of which has been completed. The long-term plan is to extend electric traction from mountainous routes to overloaded trunk links in eastern plains. The system adopted is 25 kilovolts 50 hertz. At substations, which are of 10, 15 or 20 megavoltampere rating according to location, capacitors are installed to raise the power factor. Electronic remote control for substation supervision awaits development. Both Dalian Electrical Machinery Plant and Tianjin Electric Locomotive Factory produce electric locomotives. The latest model, the Shaoshan SS-3 from Tianjin, has an hourly rating of 4 800 kilowatt and maximum speed of 95 kilometres per hour. Provided with single-arm pantographs, it employs silicon rectifiers with tapchangers but uses thyristors to regulate the excitation of its dc traction motors. Electromagnetic interference with wayside telecommunication cricuits is severe, and intense research is going on to overcome this undesirable feature.

Similarly, advances are urgently sought in the design of a new class of locomotives for service on flat regions. Before electric locomotive manufacturing got started in the country a few units were purchased from France, such imports (and exports) being handled by the National Railways Technical Equipment Corporation. However, complete loco nits are not expected to be imported anymore. The Dongfeng diesel-electric and Dongfanghong diesel-hydraulic series, made by the Central China Locomotives and Rolling Stocks Plant at Changsha and elsewhere, are testimonial to the technical level attained by the country. The Dongfeng DF-4 model, which serves as the workhorse of her diesel fleet, is furnished with the 16240-Z engine (3 300 horsepower), TQFR-3000 generator, GTF-4800 silicon rectifier set and ZQDR-410 traction in Co-Co axle configuration. Its maximum speed is 120 kilometres per hour and akin versions have been exported to some third world countries. Steam units such as the HP and the QJ 2-10-2 types, both around 3 500 horsepower and 80 kilometres per hour maximum, are likewise reliable designs that will remain in production at Datong and elsewhere for some years to come. With coal and manpower in plentiful availability, the government has rightly decided not to replace steam by diesel motive power, in disregard of the many advantages of the latter.

At present gondola and hopper cars comprise the largest single group of freight rolling stock, followed by tank cars for moving oil, and small numbers of special-purpose railcars such as refrigeration cars for food transportation and flat cars for missiles. Passenger coaches are large, roomy and comfortable (unless over-crowded!) but still unnecessarily heavy: new ones will incorporate lighter bodyshells with modular components and utilize more aluminium alloys and plastics in their construction. Airconditioned coaches are in production and plying the Beijing-Shanghai, Beijing-Guangzhou and Guangzhou-Xianggang routes. Airtight ones with pressurized cabins are on the drawing board for future use on the Xining to Lhasa line. They will be hauled by diesel locomotives with two stages of turbocharging to overcome the low atmospheric pressure.

In China railway control and signalling are still manual to a considerable degree. Some 36 000 kilometres of line have been equipped with semi-automatic relay block systems but only about 1 200 stations with electrical interlocking. About 8 400 kilometres are provided with track-to-train radio communication; a few other sections have microwave communication. A computer-aided traffic control system is the focus of research and planning.

To end this section we take a brief look at city transportation on rails. The sole urban railway system so far constructed is the Beijing underground. Work commenced in 1965 on the first line of the system; this line consisting of 17 stations began operation on 1 October 1969. The second line of 12 stations, built since 1973, has only just become ready; the delay arose out of resources being diverted to aid Korea in building the Pyongyang Metro. Both lines have had to be constructed by the cut-and-cover method. All equipment on them are home made. Electric trains are supplied jointly by Changchun Rolling Stock Works in Jilin province and Xiangtang Electrical Machinery Plant in Hunan. The former, the biggest of its kind in China, is the main manufacturer of lightweight passenger cars. The latter has been the first in the country to produce electric locomotives (the 6-Y-1). The underground locos feed on 750 V dc and mostly have a maximum traction speed of 80 kilometres per hour. A Chinese computer is installed at the control centre at Xizhimen Station, for monitoring operating conditions and issuing mandatory commands that will automatically slow down or stop trains in contingency cases.

8.2 Aeronautics

China did not seek any strong presence in the international civil aviation scene until the early 1970's, when she began acquiring air traffic rights in many more cities abroad and importing passenger jets from the West. Her emergence is being taken seriously by the worldwide airline community. Represented by the body referred to as the Civil Aviation Administration of China, she now operates eighteen international routes that amount to 140 000 kilometres and are served by Boeing 747SP's purchased in 1979. For domestic routes, which number 173 and total 210 000 kilometres, her fleet contains about twenty varieties including the Boeing 707-320, Hawker Siddeley Trident 2E and Ilyushin 62, besides self-made types. The domestic fleet was augmented by ten Boeing 737-200s and a 747 Combi delivered in 1983. In contrast, in the military sphere most of the aircrafts are, by both choice and necessity, indigenous in production though none is yet entirely so in basic design. This self-reliance is not a hypocritic act of making a virtue out of imposition, as shown by the recent Chinese decision not to buy Harriers which the UK had offered to sell. The only import move is the one under consideration about the purchase of French Mirage 2000, and concerning this deal the Chinese are going slow. Thus, the country's aeronautic industry caters chiefly for defence requirements. It is under the control of the Third Ministry of Machine Building, which since 1982 has been renamed the Ministry of Aviation Industry. This government body has close links with, financially, the Ministry of National Defence and, in technical matters, the Science and Technology Commission for National Defence.

As mentioned in the introduction to this chapter, inland air traffic is insignificant in volume compared to rail transport. A third of civil aviation flying time is devoted to activities which cannot be easily conducted otherwise, such as the application of fertilizers over wide areas, aerial afforestation, forestry patrol and fire fighting, fish shoal monitoring, airborne geophysical surveying, rain cloud seeding, and ferrying of essential people or cargo to remote spots. China produces a range of machines suitable for such purposes, some of which are being offered on the international market. The Y-5 ('Yun' being Chinese for 'transport') is a light piston biplane capable of flying at low altitudes and accommodating 11 passengers or, as an ambulance craft, 6 beds attended by 2 medical attendants. The Y-11T1 is a 17-seater. Manufactured at Harbin, this export version of the Y-11 has twin PT6A-10 turboprops from Pratt & Whitney but, with the additional exceptions of some avionic instruments such as a Doppler radar, all other components are home-supplied. The Y-7, powered by a pair of Pratt & Whitney PW 100/2 turboprops, has a wing span of 29.2 metres, maximum take-off weight of 22 tonnes, cruising speed of 480 kilometres per hour and a range of 1 900 kilometres. It can land on rough, soft runways and has short takeoff, so that if necessary it may work on apron space only. Its special design characteristics include extra lift to enable it to take off from locations where the air is hot or thin, for adaptability to the extreme geographical conditions encountered in the vast country that China is. Its flight tests were finished in mid 1982 and serial production has commenced by early 1983.

Like the Y-7, the medium transport Y-8 is expected to be in full-scale production soon. Specifications are: four Chinese made WJ-6 turboprops rated at 4 250 horsepower each, wing span 38 metres, tail height 11.16 metres, maximum take-off weight 61 tonnes, cruising speed 520 kilometres per hour, range 5 500 kilometres and minimum runway lengths required 1 270 metres (take-off) and 1 050 metres (landing). The Y-10 is the largest indigenous model, which is also the first self-designed jet. It uses P & W JT-3D/7 turbofan engines as does the Boeing 707, to which it resembles in overall dimensions, being 128.8 metres in length, 40.5 metres in tail height, 126.7 metres in wing span and 19.8 metres in wheel base. A few pre-series Y-10s have been assembled in Shanghai and long-distance test flights successfully performed from the end of 1981 onwards. Modified versions envisaged are a freighter, troop transport and an AWACS-type early warning aircraft for the military.

The backbone of the Chinese fighter force consists of the J-6 and Q-5 families. In older literature, following Western custom, they have been referred to as F-6 and 9 respectively; the Chinese character 'Jien' stands for 'to destroy' and 'Qiang' for strong. Both are suitable for active service only in good weather during daylight hours. In performance the transonic J-6 is comparable to the MiG 19SF day intercepter. China started assembling MiG 19s, initially relying on assistance and the importation of components from the USSR; four months after Soviet technicians left with all blueprints in August 1960, the first machine took to the air. Completely local production commenced in 1964 and, equally remarkably, will probably not stop for some time to come. Designated the J-6, the Chinese version has seen a number of improvements during the two decades, but is still easily recognizable by its bifurcated nose intake. Its engines comprise two Wopen WP-6A (the equivalent of Trumanski RD-9B-811) turbojet units ('Wo-pen' being the Chinese for 'turbo-jet'), each providing a maximum dry thrust of 25.5 kN, or 31.9 kN with afterburning. The J-6 Xin ('Xin' signifying 'new'), an updated version incorporating a slim pointed radome in the nose intake, comparable in performance to the MiG 19PF limited all-weather fighter. In overall terms the J-6 family are innately robust warplanes of laudable design by the aerodynamic standards appropriate to jets of the second generation, with armaments and electronic equipment in keeping with the period of genesis namely the early 1960s.

In 1968 the first J-7s came out. These were modelled on imported MiG-21Fs but made without access to original production toolings and engineering drawings, and subsequently improved to match the MiG 21PF in performance. Relatively few of these third-generation jets have seen service, but they have been sought after in the international market. They are judged to be relatively easy to fly and to maintain, in contrast to but in favourable comparison with more sophisticated types from the West which suffer from technological overkill. The only all-weather fighter in the Chinese Air Force is the Finback swing-wing J-8. According to American satellite intelligence test flights of its prototypes have been repeatedly interrupted and it has yet to attain full operational status. Both the J-7 and the J-8 are powered by RD-11 turbo-jets, and have concentric head air intakes to permit the installation of long-range radars in the cones. The distinguishing feature of Q-5 warplanes are their twin lateral air intakes. These supersonic ground attack fighters have been in field service from 1970 onwards, although they number far less than J-6s. Their mainstream model has two enlarged WP-6 engines; since 1976 when the Rolls-Royce RB 168-25R Spey entered production under licences in the country, variants of them have been turned out to use the Spey turbofan.

Until the beginning of the past decade, the front-line bombers of the Chinese Air Force were dominated by Tupolev Tu 4s. These long-range heavy bombers have Shvetsov ASh-73TK radial engines and cantilever wings and tail units. One of them functioned as the carrier in China's second test of nuclear weapon, a fission bomb, on 14 May 1965. Being too slow they are now expected to play a role only in reconnaissance and electronic support and countermeasures. Bombers were manufactured domestically from the mid 1960s onwards. The first to roll off the production line was the H-5 ('Hong' in Chinese meaning 'to knock'). A direct descendent of the Soviet Il-28, this type of light bomber has not yet been phased out of production. The Il-16 medium bomber was imported in 1960 without blueprints and tooling; the Chinese imitation, called H-6, was ready eight years later at Shengyang Aircraft Factory. The assembly work has since been transferred to Xian, after a suspension during 1972–77. The resumption of the work seemed to have resulted from a decision not to evolve newer types in the H-series. Capable of delivering nuclear devices of miniature to moderate sizes, the H-6 now numbers about a hundred. It is likely that China plans to rely on ballistic missiles as the nuclear deterrence force (for which see following section) but has no active or re-activated programme of building up the strategic bomber arm.

A separate air fleet for the navy has existed since the Korean War. Its main elements are some 700 Tu 2s (adapted to be torpedo carrying), J-6's and even older J-5's. Shipboard aircraft deployment capability was acquired in 1978 when several naval vessels were refitted with helicopter flight decks. It is rumoured that China has been considering the construction of a carrier for rotary-wing, VSTOL or perhaps even fixed-wing non-VSTOL aircrafts. Military helicopters that are made in the country include the Z-6 ('Zi': vertical) model, which resembles the Soviet M4.

The aircraft factory at Shengyang was the earliest to exist and remains one of the biggest in the country. The J-6, Q-5, Mil Mi-4 helicopter and engines like the RD-98 all come from there. In the last few years an increasing number of aero-engines have seen ground applications, to water injection in oilfields for example. The plant at Harbin also has a long history. Among its non-civilian products are the H-5 and AIM-9B missiles with which the J-6 is equipped; militarily oriented projects it is conducting include the trial productions of an antisubmarine seaplane and the Z-6, a helicopter gun platform of indigenous design. The J-7, J-8 and engines such as the Sprey used to be made at Xian. Nanchang has a factory that builds the Q-5 attack aircraft, as Shengyang does. Other large plants are located at Chengdu, Hangzhou, Shijiazhuang, Tianjin and Beijing, as well as Shanghai which serves also as the main supply centre for avionic instruments. The workmanship at these and smaller aircraft factories is uniformly high. Under their arrangement with Rolls Royce the Chinese sold some of the Spey engines they manufactured back to the UK. The recent order by McDonnell Douglas for airframe components is another recognition of their laudable standard. The import and export of parts and accessories are negotiated by the China Air Materials Company, whereas those of complete aircraft fall under the terms of reference of the National Aero-Technology Import and Export Corporation. In the last few years American passenger jets and Bell helicopters have been purchased. Regarding technology transfers from abroad, however, the only cases since the experience with Russia are the Rolls Royce deal mentioned before and a licensing agreement signed in 1980 with France to produce Aerospatiale Dauphin SA 375-N helicopters. None of these hardwares has become a visible part of China's arsenal. Apparently she wants to be dependent on her own comprehensive efforts in research and development, which potentially can provide her with an integral system of aerial defence unburdened by individually incompatible (and anyway relatively outdated) units bought from several foreign countries.

Jet propulsion was one of the twleve topics picked for research in the 1956 Twelve-Year Plan and has continued to receive emphasis since. Nevertheless, Chinese expertise in the design of turbofan engines and especially of high-performance turbojets cannot be judged by the same standard appropriate to the workmanship in manufacturing. The limitation apparently stems mostly from a difficulty in bridging the gap between theoretical knowledge and practical skill. The Chinese have not yet caught up with the

West in the technology of air compressors and even more so concerning high-temperature turbines. Similarly, their weakness in designing transonic aerodynamic structures may be traced to theirs in translating wind-tunnel models into full-scale prototypes. A common deficiency in experimental facilities relates to instrumentation for aerodynamic measurement and data processing as well as mainframe computers for numerical analysis.

In contrast, problems with constructional materials for aircraft body and propulsion unit are light. National production of aluminium is low in volume but has the potential to expand rapidly. The technologies of electric and arc vacuum furnaces have been mastered for the making of nickel-chromium and other refractory alloys. Titanium output has already been stepped up (Chapter 6.5) and should easily satisfy the demand by the jet (and rocket) industry. With respect to fuel, the country is self-sufficient in kerosene, although little is exported. Indigenous innovations include the development of a range of solid lubricants serving as anti-erosion additives, by the Lanzhou Institute of Chemical Physics which we have met in Section 2 of the previous chapter.

Not much is known about the research centres maintained by the Ministry of Aviation Industry. It joins the education ministry in exercising dual leadership over a number of aeronautical colleges. Colleges that offer four-year courses to post-secondary students have been established at Beijing, Xian and Nanjing. More recently, colleges that provide three-year courses have been set up at Shenyang, Zhengzhou and Nanchang. The one in Beijing (BAC) is the earliest to be set up, in 1952, as the amalgamation of eight of eleven aeronautical departments then existing in the country. The BAC has departments in aircraft engineering (aerodynamics and structures), aero-engines, aeronautical automation, electronics, computers, aviation supplies and aeronautical economics. For research there are laboratories of low-speed wind tunnels, high-speed wind tunnels, engine construction, engine combustion, metallurgy, aerodynamic heating, and flight simulation. Each of the low-speed wind tunnels has a cross section of 1.5 metres square, and a maximum flow speed of 40 metres per second at a Reynolds number of 500 000. There are two high-speed wind tunnels, powered by compressors of the blowdown type capable of Mach number 1.5 and 3.0 respectively. From BAC's drawing board have come several successful designs, the most well known of which is the Beijing Number 1, a scale-down based on the Soviet Yak-16 passenger transport and powered by twin Ivchenko AI-14R engines.

The Northwest Polytechnic University is under the sole jurisdiction of the Ministry of Education but it has an aircraft engineering department. The Department supports an Institute of Aircraft Structural Mechanics and Strength Analysis, Aircraft Design Teaching and Research Group, and Aerodynamics Research Group. Among the facilities they enjoy are a subsonic, a 2-dimensional transonic and a transonic-supersonic wind tunnel. The Department of Engineering Mechanics in Qinghua University has a strong group studying high-speed, high-temperature gas dynamics. The group won a state award in science as early as 1956 for work on gas turbines of airplanes.

Within the Academy of Sciences, the Institute of Engineering Thermophysics in Beijing is a research centre for gas turbines. Composed of four divisions carrying the respective titles of Engineering Thermodynamics, Heat and Mass Transfers and Combustion, Aero-thermodynamics of Turbomachinery and Aero-thermodynamics of Heat Engines, the Institute has contributed towards the mastery of supercharging, film cooling and high-altitude re-ignition techniques. An illustrative piece of current work is the development of a new method, based on the mean stream surface theory, for designing the outer and inner surfaces of an aerodynamic turbomachine. In-house experimental installations include a 250-kilowatt and an 850-kilowatt centrifugal compressor test rigs, a more modern 3 000-kilowatt axial compressor test rig, and subsonic and supersonic cascade wind tunnels. A doppler-effect anemometer and a Schlieren photographic facility for shock wave investigations are examples of recent equipment acquisition. There is also a laboratory for experiments on heat pipe.

The Institute got its present name on becoming independent in 1980, before when it existed as the Laboratory of Power Engineering in the Institute of Mechanics. Founded in 1956, the latter has been prominent for work on aerospace applications of jet propulsion and other subjects. Involvement with the aeronautical industry started with participation in the development, testing and evaluation of the RD-9B turbojet engine for the J-6. There are now eleven laboratories which specialize respectively in: basic research in mechanics, fatigue and solid mechanics, mechanical properties of materials, physical mechanics, measurements in mechanics, subsonic-transonic-supersonic aerodynamics, high-speed aerodynamics, mechanics of explosion, shock tube technology, plasma dynamics and magnetodynamics, and gas dynamic laser. Among the many on-going programmes one is on turboblade cooling; significant analytical and experimental results have been obtained on turbulent heat transfer under the condition of coolant injection. Test facilities for classical work on gas dynamics include a blowdown supersonic and an arc heated wind tunnel, a hugh shock tube, and free-flight ballistic ranges for high-speed, high-temperature and high-pressure experiments. A new direction of research is geophysical fluid dynamics (see p 96). Many of the projects pertain to the application of gas and plasma dynamics to missile and rocket carrier technologies. Some concerning spacecrafts per se are also undertaken, such as one on temperature effects in artificial satellites. Little information is available about them, however, except that they have met with quick success as told by the fast pace of Chinese progress in rocketry.

8.3   Rocketry and Satellite Technology

Guided missiles and artificial satellites are both the responsibilities of the Ministry of Space Industry, called until recently the Seventh Machine Building Ministry. Production of rocket propelled weapon systems for the tactical theatre is, however, controlled by the Ministry of Ordnance Industry (Fifth Machine Building, formerly). With respect to such weapon systems Chinese technology still lacks sophistication. For example, the ship-to-ship missiles installed on several types of frigates and fast attack missile crafts modelled on the Soviet Osa and the older Komar class, as well as the Luda- and the Anshan-class destroyers (ex-Kotlin and ex-Gordy respectively), are copied SS-N-2 missiles. The type of missile-tracking radar in service is of the 'Square Tie' design evolved by the Russians in the 1960s. A newer SS-N-3 missile may have been indigenously developed but, if this is true, nevertheless the batch-production stage has not been reached. There has been a talk with British Aerospace for fitting nine Luda-class boats with Sea Dart missiles but the plan is now cancelled. The negotiation represented the first arms deal directly with the Western block, and the decision reflects a renewed emphasis on self-reliance in military technology.

In the strategic theatre, the latest advance is marked by the trial launching of a carrier rocket from underwater in October 1982. The analysis of probable launch site and target area location shows that the rocket tested was a medium range ballistic missile (MRBM) with a range of approximately 1 200 kilometres. Western sources suggest that this MRBM has a solid-fuelled two-stage motor bearing similarities to the Polaris. It has strategic significance as China has built up a fleet of ballistic missile submarines, consisting not only the 'Golf' class which is conventionally powered, but also a nuclear type, codenamed by NATO the 'Han' class. The latter is capable of global mobility, long endurance at sea and close approach to enemy coasts. No production lines appear to have been set up for this SSBN, however.

On land, the first ballistic missile was test-fired in October 1966. Designated the CSS-1, it is a mobile MRBM probably 0.9 metres in diameter and 6 metres high, with a range estimated variously from 1 200 to 1 800 kilometres and a circular error probability perhaps as large as 4 kilometres. Using liquid propellant and the usual inertial (and maybe radio command as well) guidance system, it is modelled on either the Soviet SS-3 Shyster or the SS-4 Sandal: foreign observers have not settled this argument among themselves. The CSS-2, an intermediate range ballistic missile (IRBM) that can reach a distance of 2 500 to 4 000 kilometres, was indisputably developed by China relying on her own strength. Between thirty and forty of this IRBM have been introduced into service since 1971. Tested, but not believed to have yet been deployed in significant numbers, is the CSS-3, a limited range ICBM that can travel about 5 900 kilometres. It is thought that ten launch sites are being constructed for the CSS-3, and of them two have been completed as of May 1983. Full range intercontinental ballistic missile (ICBM) capability was demonstrated on the morning of 18 May 1980 when CSS-X-4, the X signifying Experimental, was launched into the Pacific over a distance of 11 000 kilometres. This missile is said to be 200 tonnes in weight, has two liquid-fuelled stages and can carry a nuclear warhead of the 4-megatonne class.

The FB-1 is a civilian version of the CSS-X-4 that has been utilized in China's space programme over the last years. The programme has been translated into a series of feats which are remarkable for having nothing to do with Soviet or American assistance, a qualification not applicable in the case of any other fellow latecomers to the 'space club', especially when rocketry is considered. The first accomplishment dated from 1957 when sounding rockets were tested. The premier satellite, codenamed SKW-1 went up on 24 April 1970 atop a CZ-1 booster. (CZ in Chinese stands for 'Long March') It is interesting to note that the intervening period of 14 years happen to be as short as the time taken for the first successful test of atomic bombs. The SKW-1, weighing as much as 173 kilograms, stayed aloft for 24 days while bleeping to the tune of 'the East is Red'. The CZ-1 has three stages, the first two of which have liquid propellant and the third has solid fuel. Since 1970 eleven other space flights have been made. In January 1978 the first orbited package to return to earth landed correctly in the target zone. Three separate payloads were placed in orbit by a single carrier rocket in September 1981. The latest mission, in which an ejected pod with photographic capsules was brought down after five days on 14 September 1982, attested to the speedy progress the Chinese are maintaining. This last recovery task was, according to an engineer from the Ministry of Space Industry while speaking to foreign correspondents, technically more difficult than those so far attempted by the USSR and the USA.

Indeed, Chinese satellite technology, although uneven (it remains weak concerning for instance onboard computers), is on the whole on a par with or at most slightly behind the state of the art in the US. After a visit to China in 1979, a group of experts representing the American Institute of Aeronautics and Astronautics characterized the gap as variously between zero to ten years, depending on particular disciplines, and in a few areas the US may actually have been surpassed. An example of the closing gap is the Chinese current development of the hardware for a thematic mapper, designed to provide ground pictures of high resolution possibly at 50 metres. The Chinese are already at the forefront regarding orbital tracking, as evident from their successes in recovering spacecrafts, and also with respect to launch vehicles, as demonstrated in substance by their CZ-3 rocket, of which more below.

All their fourteen satellites have been launched from Jiuquan Space Centre, which has been dubbed the 'oasis of modern science' as it is situated on the southern edge of the Gobi Desert, in Gansu province. This location actually places it a stone's throw from the western end of the Great wall, monument to the ingenuity and perseverance of the Chinese people at an earlier time, but may not really be ideal because of its high latitude. Constructed in the late 1950s, it has steadily expanded since. It houses sophisticated tracking and telemetry equipment for monitoring and controlling carrier rockets during ascent. For long-distance operations there is a nationwide network, the main command base for which sits somewhere in Weinan prefecture 65 kilometres to the northwest of Xian. The coverage is extended by two 'space event ships', called Yuan Wang I and II and each displacing 17 500 tonnes. These vessels have been built at the Jianguan Yard in Shanghai and fitted with Chinese computers and navigation instruments, but carry Doppler and Sidescan radars of foreign make. The land network augmented by the ocean-going tracking platforms has proved itself capable of pinpointing the re-entry and splash-down of space probes and ballistic missiles.

The country has certainly caught up with the USSR and the USA regarding satellite-tracking camera. The Nanjing Astronomical Instruments Factory makes, besides a satellite laser range finder (in cooperation with Shanghai Observatory), the HC-1 optical film theodolite. This camera has a Schmidt telescope that gives 5×10 degree field of view. The primary mirror is 85 centimetres in diameter and the doublet correcting lens 60 centimetres in aperture; both are of high optical quality. Satellite tracking and subsequent orbital analysis are under way at all the observatories maintained by the Academy of Sciences, namely Beijing (with four observing stations), Shanghai, Purple Mountain, Yunnan (located at Kuming) and Shaanxi (at Lintong). The one named after Purple Mountain is on the outskirts of Nanjing. It was its celestial mechanics group that designed and later verified the orbit of the 'East is Red' Satellite. Its outstation in Kuming, situated on top of Phoenix Hill in the eastern suburb, was detached in 1972 to form the Yunnan Observatory. Both Purple Mountain and Yunnan Observatories each have a TQ 16 computer for tracking satellites. The TQ 16, made by China since 1975, is a 48-bit machine possessing 32 kilobytes of memory and with a speed of 100 000 operations per second. The same model is installed at Nanjing Astronomical Instruments Factory and used for ray tracing in geometrical optics.

The celestial mechanics section of Yunnan Observatory owns a SBG satellite camera with a laser ranging gauge and a two-frequency Doppler satellite measuring device. This and other sections of the Observatory have undergone tremendous growth in recent years, a main reason for which being the fact that, compared to all other stations in the country, the southern sky visible at Kuming extends eleven degrees or more further down. To the north, in Hefei, Anhui Institute of Optics and Fine Mechanics has groups working on laser range finding and atmospheric transmission of laser beams. Lasing devices and materials are also studied there, including turnable, picosecond pulsed and excimer lasers as well as Czocralski grown ruby, garnet and nonlinear crystals.

The Chinese have collected and analysed voluminous amount of data on gravitational perturbations in orbital characteristics of their satellites. They are also enthusiastic about the information that may be deducted from the altimetric measurements by Seasat, though they are not in possession of recordings of the data.

In the Academy of Sciences system, the major research establishment that deals with geodesy and earth gravimetry is the Institute of Geodesy and Geophysics (p 127). In the early days, the Institute cooperated with the State Bureau of Surveying and Mapping in producing the plan for a general gravity survey of China. It has also participated in the establishment of fundamental, high-precision controlling gravity points, and took part in drafting the plan for astro-gravimetric levelling on the basis of rectangular coordinates and in determining accurately the longitudes of fundamental astronomical points in the country. It has proposed a method of determining time with the use of the contact micrometer of universal theodolite T-4 to observe east and west stars at equal altitude. Along with the national development of space sciences, geocentric coordinates of the Chinese geodetic network are being measured. It devised a programme for calculating the 1×1 degree block. A method for calculating the disturbance potential of the upper atmosphere has been put forward. Advances have also been achieved on gravitational field evaluation with combined terrestrial and satellite gravity data.

If now satellite hardwares and launch vehicles are considered, the most important research centre is the Institute of Space Technology under the Ministry of Space Ministry. Its Space Thermophysics Division, as an example, has current projects on satellite modelling and similitude theory, surface absorptivities and emissivities, high-temperature and cryogenic heat pipes, and cryogenic thermal insulation.

The premier space project in the mind of the government at present is to place a communication satellite in geosynchronous orbit over the Indian ocean; in February 1983 it was announced that this would probably take place by early 1984. In the past the intention had been aired for buying a comsat from abroad, and a tentative agreement worked out with Messerschmitt-Boelkow-Blohm on a joint definition study. Now the agreement has been suspended and the decision, reached apparently after lengthy internal arguments, is firmly in favour of making the satellite at home. For the launching there have been talks of reserving a place in the space shuttle and awaiting the arrangement by the Americans of boosting payloads into higher orbits, but such foreign dependence is now again ruled out. In fact, a domestic launch vehicle for the job is almost ready. This rocket consists of three liquid-fuelled stages and is dubbed the CZ-3 (CZ-1 has been mentioned above; CZ-2 never became operational). Vaguely similar in appearance to the Titan, it can lift over 900 kilograms into a 36 000-kilometre orbit or a greater mass such as a skylab into lower orbits. Its upper stage uses liquid hydrogen as the propellant: the Russians have tried the same system in the early 1970s – and failed. At the Unispace 82 Conference China told fellow third world nations that they can benefit from her CZ-3 launcher service within a few years.

The country has already been practising the correction of altitudes and orientations of spacecrafts, a skill essential to the precise attainment of geostationary orbits. The correction was done with miniature solid-fuelled rockets fixed to the crafts but, in 1982, she revealed the successful testing of pulsed plasma thrusters which offer greater reliability and manoeuverability. These ion rockets have been developed at the Electric Propulsion Laboratory in the Space Systems Division of the Space Science and Technology Centre, Beijing. The Centre serves as the general coordinator of all space related research, scattered at different institutes under the Academy of Sciences, an example of which being the work on friction and lubrication in vacuum at Lanzhou Institute of Chemical Physics. The thrusters it (the Centre) has developed will also be incorporated into the orbiting astronomical observatory that it is planning; in addition, as said in Chapter 6.2, its tasks include the design of an earth resource satellite. Incidentally, one of its five divisions specializes in space sciences. However, work of such nature is more concentrated at the Institute of Space Physics located likewise in the capital. There, cosmic ray propagation as well as interactions between solar wind and magnetosphere, ionosphere and the upper atmosphere are under scrutiny. Among the experimental studies are measurements of magnetic field, electron density, energetic particles flux, medium ultraviolet radiation and X-rays, by instruments abroad satellites or sounding rockets. Model experiments on magnetic storm are also in progress. Some research on space sciences is also conducted in the Geophysics Department of Beijing University, and in the Earth and Space Sciences Department belonging to the University of Science and Technology of China at Hefei.

8.4   Telecommunication Engineering

On the ground, the Chinese have perfected their technology of transmitting to and receiving from comsats. Since 1972 they have been operating full-time half-circuits for international communication. From April 1978 till February of the following year, they experimented with domestic sending of television, facsimile and multiplexed telephone signals via the Franco-German 'Symphonie' Satellite. By 1982 ten ground stations had been erected and from June to October of that year channels were rented on Intelsat 5. Six of the stations were outfitting with equipments designed and made by the Ministry of Electronics Industry with help from some other research units: the performance of this equipment was verified to pass Intelsat standards. Among the units involved are the Space Science and Technology Centre described in the last section, and Guangzhou Institute of Electronic Technology. The latter is interesting in that it was founded in the year 1970; called Guangdong 701 Research Institute then, it took its present name after transferral from provincial to Academy control, in 1978 when the Academy of Sciences expanded by recouping and absorbing a large number of institutes. The other four ground stations tested on the Intelsat channels were imported because indigenous production of the required electronic systems has attained the technical level but not, yet, the output volume needed. Nevertheless, satellite communication will certainly continue to gain vigour as an infrastructure development to provide, for example, links with offshore oil platforms. An even more urgent desire is to broadcast educational television programmes over the vast expanse of the country. The government has been allocated two satellite communication slots at 4–6 gigahertz by the International Frequency Board.

There is another, though relatively minor, spin-off from China's space programme into the telecommunication scene. Plants fabricating silicon solar cells for satellite use since 1971 (section 5 of previous chapter) now also manufacture them as energy sources at terrestrial repeater stations in isolated areas.

From abroad, Cable & Wireless of the UK was awarded a contract in September 1982 to set up a telecommunication system between oil rigs in the South China Sea and their onshore headquarters to be accommodated in Shengzeng Special Economic Zone. Earlier in the same month, the company concluded another deal for a 1 000-kilometre microwave network that would span Guangdong province, to be built in three stages over the coming three years. Chinese expertise regarding microwave technology is less adequate than their expertise in satellite communications.

Nevertheless, the telecommunication sector should see general improvements gradually as it is, like transport, currently receiving a boost in funding. Investments earmarked for postal and telecommunication services in 1982 accounted for 1.2 per cent of state budget for capital construction, representing an 0.4 per cent increase over the average in the last five years. Moreover, the Ministry of Posts and Telecommunications would henceforth be allowed to retain for its own appropriation 90 per cent of its profits, instead of 80 per cent as in the past, and the same percentage of foreign exchange earnings, in place of the former 60 per cent. This Ministry runs the national communication (including postal) systems. However, separate (and sometimes duplicating) networks are maintained by a few other government departments, such as the Army and the Ministry of Railways.

The microwave relay system now connects nearly all the major cities but is almost nonexistent between provincial towns. A major expansion programme is in the planning stage, one of the aims being to widen the areas accessible to educational television broadcasts (the same purpose for developing satellite communication). The present system features a solid-state transceiver unit of 960 channels in capacity, the technology originating from the Academy of Posts and Telecommunications, of which more later. Remote places are often served only by high frequency radio point-to-point links. The line communication network, consisting of both open wires and cables, has extended into all towns and some villages but suffers from a lack of capacity. Open wires for carrying telephone signals are still commonly found; most accommodate 12 channels only. There is a network of 60-channel symmetrical pair cables, which was installed in the period from the early 1960s to the mid-1970s. For municipal feeders and long-distance lines multiconductor cable is more the rule. In the 1970s 1 800-channel coaxial cables were laid on the route Beijing to Shanghai and Hanzhou. In 1976 work also started on the Beijing to Guangzhou (via Wuhan) link. Totalling 2 700 kilometres in length, it is to be served by 19 manned repeater stations and 470 automatic repeaters. The section from Beijing to Wuhan consists of eight cables and that between Wuhan and Guangzhou is to comprise four. Engineering work has been finished as far as Changsha; survey work has just commenced in Guangdong and the whole line should be operational by the end of 1985. Coaxial cable systems, both underground and submarine, are manufactured by China but technically lag by about a decade behind the state of the art in the world. Some low-loss large-bandwidth systems have been imported. The submarine cable between Shanghai and Nagasaki, with a capacity of 480 two-way telephone channels, was laid after the renormalization of Sino-Japanese relationship by Fujitsu from Japan.

Telegraphy and telex service deserve particular note: the Chinese language is ideographic rather than phonetic and therefore presents an engineering problem. The present arrangement relies on a 'telegraph table' wherein each ideogram or 'logograph' (word) is assigned a unique four-digit numerical code, in which form the word is transmitted. Terminals are now available that automatically decode from electronic memories of 10 000 ideograms and incorporate fast printers operating on the electrostatic matrix dot principle. China has acquired the British technology of ink jet writing and its application to teleprinters may be in the pipelines. Facsimile transmission of Chinese texts is an alternative way of bypassing the problem mentioned above. The typical facsimile machine made by China can send a newspaper page in three minutes. The Central Meterological Bureau (Chapter 5.1) has been transmitting weather maps since 1970. Research is in progress on a laser beam colour system.

The exploitation of the laser beam as signal carrier is at an exploratory stage. A plan to build experimental lines was announced at the Electronics Industry Conference of November 1977. By now three test circuits have been installed, in Beijing, Shanghai and Wuhan. Semiconductor laser devices are studied at the Physics Department of Beijing University as well as the Institutes of Semiconductors and Physics under the Academy of Sciences. The effort is almost exclusively concentrated on the gallium arsenide-gallium aluminium arsenide varieties, particularly the double-heterojunction types. Crystal growth process is conventional liquid phase epitaxy (LPE). The Liquid Epitaxial Group in the Institute of Physics constructed, on the basis of experimental data from differential calorimetry, a detailed phase diagram of the ternary gallium aluminium arsenide alloy in the dilute aluminium region, and lasing action was achieved in the visible when one such composition replaced gallium arsenide in the injection region. Outstanding contributions to the growth of GaAs single crystals have been made by the Institute of Semiconductors, of which a fuller description can be found in section 6 below. In 1975 continuous wave (CW) emission was obtained at the Institute from gallium aluminium arsenide double-heterojunction at room temperature. Attention has subsequently shifted to the problem of junction degradation. Devices now fabricated there have a mean operative lifetime exceeding 12 000 hours when used in prototype optical fibre communication systems. Also being investigated is the 'PNPN negative-resistance laser', which effectively houses a heterojunction laser within a PNPN structure, such that light is generated when the latter switches to and holds in the high current state. Its potential application to pulse code moderation digital communication is obvious.

Basic work on low-loss optical fibres and associated compounds like face plates is current at Xian Institute of Optics and Precision Mechanics, also under the Academy. Incidentally, besides the Fibre Optics Division, other departments at the Institute are concerned with precision mechanics, applied optics, optical processing, optical materials, radioelectronics and electron optics. High-speed photography is a strong field there. Streaking and framing cameras of the image converter type, as well as the rotation mirror and the prism compensating types, have been developed to the prototype stage. In the past these cameras were bought from abroad; those employed to monitor the tests of nuclear devices had come from Eastern Europe, for instance.

Wuhan Institute of Physics, again of the Academy, has a group studying rubidium laser. Work most relevant to telecommunication is performed, however, in the division (out of six) devoted to ionosphere physics and radio propagation from long to short wave. Ionospheric storms are studied at the Geophysics Department of Beijing University. One very recent result is the construction of a theoretical model for the negative phase of a storm at middle latitude, with which the variations of the critical frequency of the F2 layer in the atmosphere can be calculated. The Institute of Electronics, in Beijing, conducts basic research on microwave propagation in the atmosphere as well as along waveguides. Interests also extend to the generation of microwave. Many technical problems relating to components such as millimetre-wave klystrons and long-lived travelling wave tubes have been solved. One of the latest designs to have emerged is a pulsed klystron operating at 250 kilovolts and rated at 15-megawatt peak power.

The Institute of Semiconductors, discussed earlier, contains a Laboratory of Microwave Devices where concerted efforts are put behind the development of oscillators and mixers. Components that have been successfully produced include ion-implanted silicon impatt diodes for generating 4-millimetre waves, and L-band planar Gunn-effect diodes having epitaxially grown N+ gallium arsenide electrodes and working in avalanche relaxation oscillation mode (this mode of operation is a Chinese innovation). Other types of microwave devices that China has developed are based on the P-N junction, metal-semiconductor contact or tunnelling effect. In the Physics Department of Nanjing University there is a section on radio physics consisting of two research groups. The first, with Microwave Antenna as its title, is interested in phased arrays, Yagi arrays fed by open waveguides, metallic obstacle horns, microwave ferrite devices, phase shifters, isolators and switches. The second, the superconducting Electronics Group, works on low-noise microwave detectors, superconducting mixers at X- and K-bands, and lead/bismuth oxide thin-film Josephson junction devices. Complete systems for millimetre- and centimetre-wave communications are designed by the neighbouring Nanjing College of Engineering. It is envisaged that provincial telephone trunk lines will make greater use of microwave technology, and millimetre-wave facilities will become more commonplace to increase the coverage of telephone and television services.

An academy has been organized by the Technical Department in the Ministry of Posts and Telecommunications. Since the establishment of its headquarters in Beijing in 1954, this Academy of Posts and Telecommunications Research has seen some growth. It now oversees nine separate institutes specializing in satellite communications, radio wave propagation, microwave communications, communication cables, long-distance telephone systems, municipal telephone systems, telegraphy, circuits, and instruments. It has an affiliated factory in Beijing, at which 960-channel microwave equipment, ultra high frequency transmitters, 100-line automatic telephone switchboards, facsimile transmitting machines etc. have been trial-produced. The Ministry also exercises jurisdiction jointly with the Education Ministry over three telecommunication engineering colleges. The largest of these is Beijing College of Posts and Telecommunications. Research supported at the College covers many aspects of transmission, propagation and reception for a wide electromagnetic spectrum.

A major undertaking of the government in the telecommunication sector over the next few years will be the improvement of the telephone service. The service is poor due to long periods of underinvestment. The chronic shortage of telephones and telephone lines in big cities that are commercial or industrial centres has now been recognized to be slowly but surely undermining the economic infrastructure. Reflecting the current boost in funding, a 35 per cent increase is planned in the number of urban telephones from its present level of two millions, within the coming three years. By the end of the period all provincial and regional capitals should enjoy automatic long-distance dialling, and electronic computers will be introduced to do the jobs of supervision and tariff recording. The computer has already made its presence felt in the telegraphy office. The first installed, at Shanghai, greatly speeds up telegram relay between places without direct connection. Another noteworthy development in computerization focuses on a project in the Computer Science Department of Fudan University in the same municipality. A group in the Optical Scanning Laboratory of the Department is investigating pattern recognition with the use of a 24-bit, 8 kilobyte machine. The work is supported by the Ministry of Posts and Telecommunications, which has in mind its eventual application to automatic letter sorting by handwritten postcodes. According to a long-term projection, a six-digit number will in due course be assigned to each post office in the 2136 counties and 3 municipalities. The first two digits are to signify the province or autonomous region, the third the postal district, the fourth the county or city, and the last two the post office that will deliver the mail.

As the reverse of computer application to communications, the management of digital signal traffic has likewise constituted a new area of research. Chinese delegates have in the last few years attended all Consultative Committee on International Telegraph and Telephone sessions on modems and data networks (as well as text transmission). Computer usage in the country being in its infancy, no immediate need exists to develop more than a handful of computer rings and networks. However, the situation is almost certain to change in five to ten years' time since, as discussed in the next section, the message has repeatedly come across that computers are encouraged to appear in all aspects of life. The technological and social impact of C+C (computer integrated with communication) will be visible in China within a decade.

8.5   Computer Science

The electronic computer industry was among the four priority growth sectors listed in the 1956 Twelve-Year Science Plan and in the early 1960s the country lagged behind the West by, it was then judged, only a few years in terms of computer hardware sophistication. The gap unfortunately widened in the ten to fifteen years that followed. At present the backwardness is even more severe with respect to peripherals and software, and to human interfacing: computers, even where available, do not always find extensive and efficient utilization in the hands of potential users. The relevance of computer power in diverse environments is not widely understood. The situation may be changing rapidly, however. The China Computer Technical Services Company has recently been set up as a promoting agent. Software houses now exist at Beijing, Shanghai, Tianjin, Shenyang, Hefei and other industrial, commercial or scientific centres. The sum of US$22 million, out of the $200 million loan negotiated with the World Bank in 1981, has been earmarked for the purchase of Honeywell computers for installation at 14 universities. These developments are significant in spirit if not in substance. Computers and their application form one of the eight research fields emphasized in the original 1978 Science Plan.

The increase of computer accessibility may be illustrated by the case of Shaanxi Agricultural By-Products Corporation at Xian, which has automated its inventory processing with ease, as the softwares were included in the sale package; in the past the factory supplying the hardwares typically did not offer consultancy in softwares. Xinjiang Institute of Physics comprises three research divisions, one of which has just completed a project on the application of minicomputers in refrigeration technology for cold storage. The Hospital of Chinese Traditional Medicine and Beijing Second Medical College are experimenting with computer-aided diagnosis and therapy. This medical expert system is being built up with the help of several technological institutions, all of them also in Beijing. They are: Beijing Number 3 Computer Factory, the Institute of Automation under the Ministry of Machine Building Industry, as well as the Institutes of Automation and of Computer Technology (of these more later) in the Academy of Sciences. In Shanghai, the Sixth People's Hospital aided by the municipal Institute of Automation is conducting a parallel programme.

An equally encouraging sign is the improvement of computer environments at the research institutions themselves. Among the more wealthy ones the case of Beijing Institute of Physics belonging to the Academy is typical. In 1980 the Institute took delivery of an IBM 370/165. With this moderately advanced machine fast time-sharing operation became possible and a distributed terminals network slowly grew. Much of the software was 'imported', but a powerful program library for crystallographic data reduction and diffraction pattern simulation has been evolved locally. Also, the Institute's Electronics Division contains a group with the job of implementing numerical methods; incidentally, most of the subroutines written are in FORTRAN, which as elsewhere enjoys an undeserved popularity among scientists. Another group in the same division has designed a versatile data logging and processing system around a minicomputer (the Chinese-made DJS 130: see below). The full complement of the hardware consists of two analogue-to-digital converters, a paper tape reader, visual display unit or teletype, magnetic tape drive, paper tape punch, line printer and graph plotter. Compilers available are ASSEMBLER, BASIC and FORTRAN IV. The system, which is mobile, accepts bipolar signals from up to five channels at a maximum speed of 100 hertz, with a minimum accuracy of 1 per cent. It has functioned in experiments on laser four-wave mixing and in conjunction with the rotating anode X-ray diffractometer.

Computers of all sizes from the mini range upwards and already installed in China number about 3 000, of which at least a tenth have been purchased from outside. However, domestic output now reaches between 500 to 800 a year; that of all types of peripheral units exceeds 75 000. Serial manufacture of computers comes under the authority of the State Bureau of Computer Industry, which employs some 75 000 staff. Amongst its major research and production units is the Beijing Number 3 Computer mentioned before.

In the Academy of Sciences system, the oldest establishment in the computer area is Beijing Institute of Computer Technology. With assistance in workmanship from Beijing Wireless Telegraph Factory, the Institute assembled the country's first prototype computer in 1958, twelve years after the world's first. This, the Model 103, was a first-generation machine with magnetic drum memory. It resembled in performance the Soviet URAL 2 or BESM, being capable of 2 000 operations per second (OPS). In 1965 the Institute designed, again, the first transistorized computer: the 109B, a 48-bit installation with 32 kilobyte core memory and a speed of 115 000 operations per second (OPS). The first integrated-circuit unit, on line in 1970 just after the high tide of the Cultural Revolution, originated from the Institute too. Designated the '111' and capable of 180 000 OPS, it used transistor transistor logic (TTL) and a 32 kilobyte memory of 48 bits.

Since 1975, which year marked the establishment of the Computer Industry Bureau, less emphasis has been placed on the making of a few prototypes incorporating up-to-date technology and more on the provision of standard lines of products in large quantities. Many types in the DJS series (as well as the DJM 300 analogue computer series) have been made as outcomes of a fruitful collaboration among Beijing Institute of Computer Technology, Beijing Number 3 Computer Factory, Tianjin Wireless Technology Research Institute, and the Department of Computer Technology and Science in Qinghua University. The DJS 260, a mainframe (64-bit) of fourth-generation large scale integration (LSI) technology, has been available since 1978 and works at 1 million OPS. Models with a speed of 5 million OPS were designed in the same year. The overall impression is that in terms of internal speed Chinese mainframes, while more than an order of magnitude slower than US machines, compare not badly with Soviet products.

The most powerful indigenous minicomputer so far identified in China is the DJS 130, which operates at 0.5 million OPS (and, as usual, with a word length of 16 bits). The first microcomputer is the DJS 050 announced in 1977. It has an 8-bit processor constructed of metal oxide semiconductors (MOS) LSI components, similar in architecture to the 8080 introduced by Intel in 1972 but which is still an industrial standard. A new series, the DJS 060, is under development by the collaborating group mentioned in the last paragraph. One of the models, the DJS 051B, is remarkable in being a set of three pieces: television screed and tuner; AM/FM turner, cassette tape deck, amplifier and loudspeaker; and the micro itself! Although appliances like the 051B are obviously intended for home use, the dawn of the age of personal computers may not yet be at hand, given the limited capacity for manufacturing micros in the country. Demand is expected to exceed supply for some years to come. A large number will be needed to drive the proliferating intelligent computer terminals: an impetus for the popularization of microcomputing is its socio-economic implication, namely the creation of a technological base for efficient decentralization. The Institute of Computer Technology concluded a contract in 1982 to export BCM 3 single-board computers to West Germany, a move representing China's entry into the world market.

The output of the Chinese computer industry include a small repertoire of peripherals, namely paper tape readers, teletypes, visual display units, magnetic tape drives, magnetic drum units, line printers, plotters and, since 1980, graphics terminals. The design of most of them cannot be described as advanced, in most aspects including outward appearance. The paper tape readers most commonly met, however, achieve a speed of 900 bytes per second, a respectable specification, and line printers usually output 80 characters per line at rates of 400–1 600 lines a minute depending on alphanumeric character distribution in the file being copied, typical figures for a Western printer being 300–1100 lpm for 132-character lines. For some reason in the country magnetic drum is still exceedingly popular, and so is the paper tape – to the virtual exclusion of punched cards. Chinese computers as a rule are provided with only limited secondary memories: most users cannot form the habit of keeping permanent disc files. Floppy disc and winchester drives for use with micros are just beginning to be serially produced.

As in the case of telegraphy, the peculiarity of the written language places special requirements on those terminals that have to handle texts for which the alphanumeric set does not suffice. For this particular application China has made equipment of advanced designs. The '1448' Research Institute belonging to the Ministry of Electronics has just developed an input terminal, the keyboard of which contains 96 keys standing for ideogram components and capable, through the action of 17 control keys, of encoding 4 164 complete ideograms. An output terminal that can print 70 to 120 ideograms a minute was recently put into coproduction with West Germany. This 'intelligent' teleprinter, with a memory of 7 000 ideograms, is based on the decoding method invented at the Research Institute run by Shanghai Instruments and Meters Plant, and on the ink jet printing and microcomputer technologies of Olympia Company. In addition, a colour cathode-ray tube device linked with electronics for word processing has been marketed by Wuhan Industrial Machine Building Bureau.

A remarkable achievement in a related area is represented by a computer-controlled laser editing photocomposition system, which provides an extremely fast way of typesetting Chinese texts. Designed by the Computer Science Department of Beijing University, this phototypesetter relies on new methods for a high degree of information compression and a fast rate of retrieval. Different fonts of various points can be set at a speed of 100 ideograms a second.

Research on programming languages and operating systems has, like development of peripherals units, been relatively neglected; not until October 1979 was the First National Conference on Software held, two months before the Fifth on Computers (Hardwares). Since its introduction in 1963 ALGOL has been favoured and, in the modified form of the BCY dialect (see shortly below), remains the language most commonly available on Chinese mainframes. As early as 1964 groups from the Computer Science Department of Nanjing University and the East China Institute of Computing Technology jointly implemented a basic subset of ALGOL 60 on the J 501 machine. BCY was designed the following year by the Beijing Institute of Computer Technology. FORTRAN and later COBOL compilers came into use by the early 1970s and PASCAL, in 1979 FORTRAN IV takes second place after BCY in terms of popularity; first place among 'ordinary' scientists. BASIC is the usual high-level language spoken by Chinese minis and micros.

The same group at Nanjing University developed an automatic software generating system on the Model 655. In use since 1978, the system contains a sequence of languages in which each member belongs as a subset to the succeeding one. The first member, the core language, is PASCAL-like but modular, with its self-compiler written directly in the 655 machine language. Self-compilers for successive members are automatically generated in turn via bootstrapping techniques. Soon after the introduction of this 655 system, a general system programming language called the XCY was developed. Base compilers for it have been specifically implemented on the DJS 200 series, but it can be easily transported to other types of computers, thanks to the localization of machine-dependent components, namely operating modes of modules, machine-specific record types, locations in variable declarations and related components. Its chief application lies in the writing of large concurrent programs such as operating systems.

The DJS 200/XT 2 operating system, an automatic diagnostic system for complex programs in XCY, is one recent example. Designed to supervise both batch-processing and time-sharing operations, it has been the work of the Computer Science Department in Beijing University. Composed of 14 paths, it is constructed layer by layer from the virtual machine viewpoint. It comprises a kernel and six layers of modules, and represents the first of the third-generation operating system to be designed in China. The earliest first-generation system is the DJS 11 for the DJS 150 computer. Written in 1970, the DJS 11 permits only up to four user programs to be executed concurrently. Multiprogramming is as yet impossible on microcomputers. For example, the 052 FDOS floppy disc operating system for the DJS 052 is a single-user, single-task system which has just become available. However, with the appearance of the powerful 060 series more sophisticated operating systems are envisaged.

Research in algorithm analysis and other mathematical studies have gathered momentum in recent years. As an illustration, work on complexity theory aimed at the minimization problem connected to the numbers of steps required for specific types of computation, is being taken up at the Institute of Computer Technology.

This institute was founded in 1956 and, as remarked before, qualifies as the oldest computer research establishment in the country. It has designed or helped to design a number of computers and peripheral units, some of which have been noted above. Its ten divisions are devoted to computer architecture, central processor and basic circuits, memory technology, peripheral units, power supply, packaging techniques, computer automation design, software and basic theories, computer applications, and instrumentation. It operates a large factory where prototypes are built and experimentally run for engineering and software developments.

The Institute of Automation is, again, located in Beijing and engaged in designing computers and accessories, besides other activities. It specializes in remote terminal displays; among its computer products, the DJS 4 is a third-generation machine still finding applications, in the control of ammonia synthesis at many middle-scale chemical and fertilizer plants for example. Its participation in a project on computer-aided medical diagnosis has been mentioned previously. Peripheral units are also under the development at Harbin Institute of Precision Instruments, newly established in 1978.

Shenyang Institute of Computer Technology has functioned since 1958. Its attentions are directed towards minicomputers and, since a few years ago, micros. Organizationally it consists of nine laboratories: system and logic designs, information storage, peripheral devices, power supply, system software, computer-aided design and application softwares, computational mathematics, microcomputers, and information analysis. A factory where pilot production takes place is attached. The mini currently being designed has the designation SJ 55/40.

Some minis are manufactured at Chengdu Scientific Instrument Factory which was set up in 1959. The Electronic Computer section there makes the Great Wall 203 with associated interfaces for environmental protection applications. Among the other three sections, the one with the title 'Electronic Measuring Techniques' has developed precision temperature controllers, thermographs, a waveform digitizer, etc. Outputs from the Optical Instrument section include stereo microscopes (and overhead projectors); some of the work at Remote Sensing Techniques has already been discussed on p 126.

Chengdu Institute of Computer Application came into existence in 1966, then as the Computer Technology Institute under the dual leadership of Fifth Machine Building Industry and National Defence Commission. It was placed under the sole jurisdiction of the Academy of Sciences in 1977 and rechristened Chengdu Computer Station, until 1981 when the present name was assumed. Its emphasis of research has shifted to software. Contributions in the past several years include the DJS 6 computer ALGOL 60 compiler, the DJS 100 series FORTRAN operating program base, a single-base algorithm language interpretative program, a real-time system for ballot processing, computational programs for planometric block triangulation utilized in horizontal control survey, and the implementation of a submatrix method for solving definite high-order symmetric linear equations. Examples of work in progress are the processing of an environmental information data base for the Ertan-Dukou region and generating executive software for microwave network analysis.

Beijing Computer Centre, opened in 1977, has as its major task the study of numerical methods and non-numerical techniques, aside from regular provision of computer time to other institutes in the Academy. The research on numerical analysis is most active. Some projects recently completed relate to mathematical physics particularly gas dynamics, the solution of nonlinear evolution equations, and adjustment computation on China's astrogeodesic net. Also in Beijing, the Institutes of both Mathematics and Applied Mathematics support work concerning the use of computers. One of the ten divisions in the former is assigned to study hierarchical software and data base. A group in the latter has interests on syllogistic deduction of mathematical formulae by computers.

8.6 Electronics

The Ministry of Electronics Industry, known as the Fourth Machine Building until 1982, carries some responsibility for manufacturing electronic apparatuses, such as the Chinese-character keyboard terminal mentioned in the last section, as well as consumer items. However, its primary function is to supply electronic components to plants under other government departments like the Ministry of Space Industry, or Posts and Telecommunications, and the Bureau of Computer Industry. From the late 1970s onwards China has taken her position as one of the ten biggest electronics producers in the league of nations. Her output already topped US$2 000 million in 1976, which year witnessed her overtaking of Japan as the second largest maker of radios. The expansion in production volume during recent years has continued to be fast in the electronics sector, which is not energy-intensive. In terms of quality, on the other hand, the level of her electronic technology is reckoned to be five to ten years behind the world's state of the art. The problem here concerns not so much research as factory production.

A main reason for this gap is the relative neglect of the sector by economic planners up to the early 1970s. The neglect stemmed from political sentiments that the steel and machine building industries should take precedence, because of their indispensability to the construction of a broad industrial base that did not exclude small plants dispersed over rural areas. In contrast, sophisticated electronics are more applicable to large enterprises and thus centralization, and to automation, which displaces employment by investment. These misgivings no longer apply today. The essential role of electronics in avionics, space and computer industries is freely recognized. Moreover, the advent of options such as the micros has made electronics more of a soft technology.

At present major efforts are being spent on solid-state digital electronics. According to the 1978 Science Plan, China 'should lose no time in solving the scientific and technical problems in the industrial production of large-scale integrated circuits, and make a breakthrough in the technology of very-large-scale integrated circuits.' The discussion in this section will therefore start on the topic of integrated circuits (IC).

The Academy of Science runs eight factories, two of which undertake trial production of ICs. Factory '109' in Beijing, operated since 1958, develops and manufactures medium scale integration (MSI) bipolar high-speed TTL and emitter coupled logic (ECL) circuits. In 1980 a production line for large scale integration (LSI) circuits was added for the purpose of quality and cost problems evaluation. Xinxiang Semiconductor Device Factory was originally the Xinxiang branch of the factory which then gained independence. Besides TTL MSI circuits, high-frequency low-noise transistors for communication equipments are also made there.

Shanghai Radio Factory Number 14 under the Ministry of Electronics is the country's leading mass production unit for ICs. This plant commenced manufacturing transistors in 1972 and a year later its output also included ICs. It was visited in 1977 by an American delegation representing the Institute of Electrical and Electronics Engineers, which reported the silicon wafers in use were 40 millimetres in diameter though the plan was to introduce 75-millimetre ones soon. Both ultrasonic and thermal compression techniques were adopted for leads bonding. Finished chips were tested by a DJS 1 minicomputer that identified failed units. Overall MSI and LSI yield remained low at about 10 per cent and masking was done by photolithography that could not go below a line-width of 10 microns. However, innovations from the Academy's Institute of Semiconductors, of which more later, are said to have led to significant increase of the yield. Research is also underway on electron-beam lithography that may attain sub-micron resolution.

Among the factories producing solid-state devices fabrication machineries, Beijing Number One Semiconductor Equipment Plant is outstanding – though not unique – in having been set up by the masses through their own initiatives. It was started in 1958 by seven housewives to make simple weighing balances of the types found in vegetable and meat markets. In 1967, when there was a grassroot movement to master semiconductor technology 'in the backyards', it went to the Radio Electronics Department of Qinghua University for technical advice and soon began turning out diffusion furnaces. Today it is a major supplier of such furnaces, although its product line has been diversified to include items such as ultra-clean laboratory rooms and transistorized cardiographs.

Qinghua University is one of the important centres of research on IC fabrication. The Radio Electronics Department just mentioned has an Integrated Circuits Laboratory that designs, among other things, laser-guided step-and-repeat cameras for mask generation. It also produces on a small scale JK flip-flops, complementary metal oxide semiconductors (CMOS) logic circuits and LSI components like 1 024-bit MOS memories. Its current R&D efforts are channelled towards silicon-gate and N-channel circuits. LSI layout designs are prepared by the neighbouring Department of Computer Technology and Science with the aid of computer-aided design (CAD) procedures. Another university that conducts research on device fabrication is the Xian Polytechnic, the Electronic Engineering Department of which is looking at integration-injection logic circuits, and the substitution of silicon tetranitride by alpha-alumina as a material for MNDS memory devices.

Within the Academy system, Chengdu Institute of Optics and Electronics has work on precision optics including photo-lithography. The Institutes of Metallurgy and of Semiconductors are the two major centres for IC technology. The former is in Shanghai; its bipolar group has produced an ECL 1 024-bit random access memory (RAM) with an address access time of 28 nanoseconds. Its metal oxide-semiconductor group is the originator of a FAMOS 8 192-bit electrically programmable read only memory (ROM). Another group there has been responsible for the microprocessor used in the DJS 050 minicomputer. Its auxiliary workshop has constructed an infrared mask alignment machine, ion miller, and ion implanter having a range of ion energy from 22 to 200 kiloelectronvolts with possible extension to 400 kiloelectronvolts. (The biggest in China can go to 600 kiloelectronvolts.) Incidentally, it has also designed and built an Auger electron spectrometer with an energy resolution of 0.3 per cent, of which good use has been made for basic research in solid-state physics. There are strong links with many semiconductor factories in Shanghai and with the Institute of Computer Science in Beijing.

The Institute of Semiconductors is rather unusual in being located within the city centre of Beijing. It has designed a computer, the BXJ 1 with a speed of 100 000 OPS and an 18 kilobyte memory of 24 bits, for the control of mask making. One of its significant contributions is the 'building block' approach to LSI/VLSI mask pattern generation. In this method, which was also independently invented by the Japanese, interchangeable patterns of basic units that are frequently encountered are catalogued. They can be positioned to within 4 microns and pieced together to form the desired mask: in this way the number of exposure instructions is greatly reduced. The technique has successfully been applied to the production of 1 024-bit MOS shift registers. Another contribution is the development of DYL circuits, in which a new 'linear AND-OR gate' serves as the basic components. ICs consisting exclusively of this kind of gates can be fabricated without epitaxy and isolation. MSI circuits can be designed to achieve a delay time per gate of 1 nanosecond even when conventional processing is employed, ie with line width as much as 10 microns and alignment allowance 5 microns. Since 1978 a wide variety of DYL circuits has been made at the Institute. Experience confirms the advantages of these circuits in speed, ease of production and high yield.

The Institute is prominent in research on solid-state devices in general. Its activities in microwave semiconductor electronics have been outlined in section 4. It was the first to realize the planar process and to fabricate silicon planar transistors in the country; for the work it was bestowed a Class I prize by the National Committee of Science and Technology in 1964. Shortly afterwards it produced the first silicon MOS transistors. These advances paved the way for the acquisition of the technology of IC, besides being on their own essential to that of discrete components. Next to silicon the most important material it, and most other places in China, use in making discrete components is gallium arsenide. Basic materials research has been carried out at the Institute since its early years. In 1962 the first single crystal of gallium arsenide was grown by the horizontal Bridgman method and, by the late 1960s, crystals that gave positive temperature coefficient for resistivity, indicating negligible concentration of deep-level impurity trapping centres, were obtained by a special annealing method developed in-house. Tellurium- and silicon-doped crystals of dislocation densities below 0.1 per cubic millimetre were prepared in 1974. One of the recent accomplishments is the reproducible growth, by either *liquid* phase epitaxy (VPE) or LPE with temperature gradient, of epitaxial gallium arsenide in which electron mobility at 77 Kelvin reaches 210 millimetres per second per volts per millimetre. Other III-V compounds being examined are gallium phosphide and indium phosphide, growth in a special crystal puller that can withstand a pressure of 100 bars. Also there is interest in some amorphous oxides and chalcogenides from the point of view of possible exploitation in switching and memory devices.

Basic solid-state physics is studied in many places. Any discourse that has the pretension of offering at least an overview needs to be lengthy, and will not be attempted here since the subject does not strictly speaking belong to electronics. However, a few examples may be instructive. The Physics Department at Zhongshan University in Guangzhou has a group devoted to electron energy band structures. The Department at Fudan, Shanghai, is involved in surface physics, some of the on-going projects being experimental studies on silicon with the help of soft X-ray spectroscopy and on gallium arsenide with LEED. An unusual piece of work, performed in collaboration with the Zhongshan Hospital affiliated to the Shanghai First Medical College, relates to the fluorescence emission spectroscopy of organic substances present on the surface of the human tongue, the end being to develop a diagnostic aid.

Nearly half of the research divisions in the Institute of Physics under the Academy specialize in areas related to solid-state electronics. The Surface Physics division built the country's first molecular beam epitaxy system in 1979 with technical assistance from Shanghai Scientific Instrument Factory. The system incorporates an Auger electron spectroscope of energy range 1–3 kiloelectronvolt and resolution 0.5 per cent, a 10–50 kiloelectronvolt HEED apparatus made by the Institute's Electronics division, and a 1-150/10-400 a.m.u. quadrupole mass spectrometer supplied by Beijing Scientific Instrument Factory (p 89). Epitaxial films of monocrystalline gallium arsenide have been successfully grown in the system: their quality confirms its equality in performance with the best elsewhere. Hardwares recently designed by the Electronics division include, besides the HEED apparatus and the data logger described in the last section, circuits for energy loss spectrometer and microprocessor-based Mössbauer spectrometer. The latter instrument is being applied to the study of ion substitution and heat treatment effects on the microscopic properties of various materials, by a group in the Magnetism division. This division is also doing research on the formation and the collapse of Bloch lines in magnetic bubbles and on the mechanisms of hard bubble suppression by ion implantation. Finally, the defect structure of float zone (FZ) silicon crystals grown in hydrogen is among the topics of interest to the Crystallography division.

Gallium arsenide is likewise the most studied material after silicon at the Institute of Metallurgy. There, gallium arsenide single crystals are prepared with the Czochralski technique for FET, infrared light emitting diode, and Schottky diode, a well-known example of which is the 8GC switching diode. VPE has been employed to grow gallium arsenide epilayers 0.5 microns in thickness for use by a collaborating institute in making infrared detectors. Visible light emitting diodes consisting of doped gallium phosphide are in pilot production. Research has just begun on liquid crystal displays.

Liquid crystal display (LCD) is also under scrutiny at the Institute of Organic Chemistry located in the same municipality. Using the electro-optical effect in the storage mode of cholesteric-nematic transition, sequential array displays have been developed and incorporated in a page composer of 32×32 bits, in 1979. The following year an organic light valve was constructed; similar light valves with organic conductors are under investigation. Another study in progress concerns colour LCD of the negative type that exploits the guest-host effect. Colour LCD is again the focus of attention at the Radio Electronics Department of Qinghua University, which in 1977 designed one of China's first digital voltmeter with LCD. The Department cooperated with Beijing Institute of Glasses to produce, in 1976, a flat television screen that relies on cholesteristics as light valve. At the Institute of Physics, a liquid-crystal bistable electro-optical modulator which may serve as a photonic logic gate was fabricated in 1980 by the Laser division. Some Chinese achievements in medical electronic applications of LCD have already been described under Chapter 3.5.

The Laser division of the Institute of Physics, the Laser Beam Originating Devices Laboratory of the Institute of Semiconductors, the Physics Department of Beijing University and a multitude of smaller units are all appropriating much resources to research on semiconductor lasers of the injection type and based on gallium arsenide/gallium aluminium arsenide materials. For a description of the work see section 4. There is apparently much duplication of efforts. Research on lasers in general is widespread in China; its popularity has been attributed by some to be a 'beehive' ('bandwagon') phenomenon.

9 Science and Technology in Taiwan

Historical circumstances have made it both necessary and instructive to treat science and technology in Taiwan under a separate chapter heading. The island province of Taiwan, less than 200 kilometres off the coast of Fujian, serves as the demarcation between the East and the South Seas geographically and, politically, between the East and the West. The reason lies in its being under a different government from that on continental China, a situation representing a protraction of the civil war of 1947–49. Its social system, indeed, is a direct continuity of what was prevalent over the whole of the country before the succession of the ruling party in 1949. In many ways, the traits of science and scientists in it today constitute a good indication of how things might have been had the series of revolutions lasting into the 1970s not swept through China.

9.1 Demographic and Economic Background

Taiwan has an area just over 36 000 square kilometres and a population of 18.3 million, concentrated on coastal plains, especially on the western coast. There are many associated islands in the territory, including the Penghu (Pescadores) group, Jinmen and Mazu, the latter two being only a few kilometres off the shore of Fujian. Tainan used to be the seat of the provincial government, which is now at Taizhong to the north, while the 'national' government sits at Taibei which is further to the north. Following the vote in the UN General Assembly taken on 25 November 1971 to expel Taiwan from the United Nations and to re-admit the People's Republic, the 'international' position of Taiwan is uncertain. With a few exceptions like South Africa and Israel, nearly all countries have ceased to have diplomatic relations with Taiwan. Trading relations are maintained with scores of countries, however.

The island is mountainous, particularly the eastern part. Nevertheless, there are fertile plains and 25 per cent of the land area is cultivated. Agriculture employs 15 per cent of the population and accounts for only 1 per cent of the GDP; as in many developing regions there is a migration of labour from the countryside to the cities. Bananas, pineapples, rice, tea, tobacco and sugar are among the main crops, with rice being the staple crop as is generally true for a southern province in China. Forestry has a huge potential as most of the interior and western parts of Taiwan, or 64 per cent of its area, are covered with forests.

Coal is on the way to being worked out as a major mineral resource (see section 8). Natural gas was discovered in 1968 north of Taizhong, and a petrochemical industry has been built up. In addition, food processing, machinery, metals and textiles are the principal industries. A fast-growing sector of the economy is the field of electronics, backed by massive inflows of capital from the USA and, more recently, Japan. At present, 40 of the 56 electronic companies with annual turnovers exceeding US$1 million are offshoots of multinationals or foreign joint ventures. Foreign money is also playing a part in the expansion of heavy industry, for example the shipyard and steelmill at Gaoxiong. Official statistics on foreign investment are noted for the modesty of their authors, but a more accurate figure can be arrived at if the total of US$3 000 million stated to be the amount of direct investment in all sectors since the early 1950s is trebled. Of this sum, about half has been from the US and a quarter from Japan.

Taiwan has a mixed economy, with the private sector dominating and, as has just been said, fuelled by investments from abroad. Economic growth has been rapid: the GDP increased on average by nearly 10 per cent during 1970–78 despite the recession of the global capitalist system. In the last few years stagnation and inflation finally cut in deeply, however. The government targets for economic growth were repeatedly revised downwards; nevertheless, in 1982 for example, the growth achieved was 6 per cent and fell far short of the official forecast of 7.5 per cent. Moreover, the per-capita GDP of US$2 200 is way behind that for the USA ($12 500), upon which Taiwan is modelled economically, and from which massive investments have been injected.

9.2   Science Policy and Organization

Science used to have low priority in Taiwan. A major reason for this is the foreign dominance in the economy that leads to a heavy reliance on the import of technology. The first semiconductor plant on the island to open (in April 1982) is an illustrative case. The factory embodies knowhow bought by United Microelectronics Inc, of which 45 per cent are under government ownership, from RCA for US$3 million and a promise of royalties. One conspicuous consequence of the general indifference to long-term research work in the past was, and is, the drain of technical manpower to the West and North America in particular. One in eight students of science and engineering leaves Taiwan within two years after graduation, while four out of five of those completing postgraduate training stay abroad or, in some cases, return to continental China instead of to the island. At the Second 'All-Nation' Science Conference held at Taibei in February 1982, the problem of brain drain featured as one of the main topics of discussion. Among the complaints aired by resident scientists at the Conference were the lack of research facilities and of participation in policy making. These misgivings against the bureaucrats were no different from those of the scientists on the mainland!

However, the last three or four years have witnessed a dramatic reversal in the intention of the government, if not yet in actual conditions. The change in the fortune of science is closely linked to a new desire to build up a domestic industry capable of producing advanced weapon systems. The underlying reason is the increasing international isolation of the regime, which therefore faces uncertainty over the future possibility of buying from former allies defensive equipments needed to maintain military superiority over the mainland. Thus, the 'state' budget for fiscal year July 1983 to June 1984 called for an expenditure of US$8 100 million, of which 42 per cent are allocated for defence and a meagre 15 per cent for education, health and culture. (On the mainland defence absorbs under 20 per cent of public expenditure.) Indeed, over the past five years in terms of the GDP, the share for defence has consistently been increasing while that for education, health and culture has been steadily decreasing. On the other hand, scientific research is planned to absorb about 1.1 per cent of the GDP, representing an 0.3 per cent rise over the mean percentage for the last five years. Of the science funds, around one-quarter is earmarked for spending by the National Science Council (see below), the rest to be disbursed to academic institutions and research centres under government ministries and public corporations.

A tangible result of the new emphasis on original research is the development, since September 1981, of a science park at Xinzhu, a town situated halfway between Taibei and Taizhong. Still unclear in effectiveness is a recruitment drive aimed at scientists of Chinese ethnic origin but educated abroad. An official report released in March 1983 recommended that government ministries should take in over the next two years 3 000 plus research staff, of which only a third can be supplied by home universities. Within the Xinzhu science park, a government middle school was started in the 1983 academic year to cater for children of 'returned' experts: the pupils are to be exempted from entrance examinations and the teaching will be in both Chinese and English.

Science policy in Taiwan is based upon various Four-Year Plans, the first of which appeared in 1968. The one promulgated by the government's Executive Council in August 1982 identified these eight priority areas for research: materials science, information technology, automation, genetic engineering, energy, laser, hepatitis prevention and therapy, and food processing. The first four have also been specified by the Ministry of Education as subjects which students will receive financial incentives to study when going abroad.

The National Science Council was formed in 1967 as successor to the National Council for Science Development as the major government agency for science and technology. It does not directly engage in research but coordinates the science activities of all ministries, being responsible for success in the implementation of government science policy. It is the most powerful body dealing with science in Taiwan and reports directly to the Cabinet; the previous National Council for Science Development did not manage to do this. The recruitment drive mentioned earlier is under its auspice. Periodicals that it publishes in the English language are the annual

Review, quarterly *Proceedings* and fortnightly *Science Bulletin*; only its *Monthly* is in Chinese.

Among independent government organs devoted to particular administrative tasks there are the following:

(1) The Council for Agricultural Planning and Development, Taibei. Called the Sino-American Joint Commission on Rural Reconstruction before 1 January 1979, when diplomatic ties were formally severed with the USA, the Council plans and funds agricultural research work. In its advisory role it provides technical assistance in rural reconstruction projects.

(2) The Industrial Research Council under the Ministry of Economic Affairs. This agency looks after most of applied R&D.

(3) The Atomic Energy Council, Taibei. It is responsible for the supervision and regulation of all atomic research establishments and nuclear installations. Regarded as being in charge of important and sensitive matters, it is placed directly under the authority of the Office of Executive Council Chairman (the equivalent of prime minister). It consists of fifteen members, who represent interested government departments, such as the Ministry of Economic Affairs, the Ministry of Education and the Ministry of Communications. It has the functions of determining the general policy for a nuclear programme, and the coordination of all activities connected with nuclear energy.

(4) The Council for International Economic Cooperation and Development, Taibei. This government body considers problems relating to the attraction of foreign investment.

The Academia Sinica is entrusted with the direction, promotion and actual conduction of research. In addition to running research institutes of its own, not all of which are science-related (eg, the Institute of American Culture), it aids and coordinates work in universities and other establishments. The Assembly of Fellows, comprising 90 elected scholars from Taiwan (but some of whom have settled abroad), delegates the authority of determining and organizing the policies and actions of the Academy to a standing council, membership of which is effectively by appointment.

9.3   Government Science and Technology

Various ministries and independent government agencies run research programmes in their own laboratories. These are outlined below.

Ministry of Communications

The Telecommunications Laboratories of the Ministry are intimately concerned with the burgeoning electronics industry in Taiwan, which, with over US$1 000 million of mostly foreign investment, is the fastest growing sector of the economy. Among research interests are the areas of communications equipment, computers, electronic switching systems, integrated and printed circuits, microwaves, radio wave research, and semiconductor materials. Owing to the Laboratories' origin in 1951 as the Radio Wave

Research Laboratories of the Ministry, atmospheric noise, radio waves, geophysical work, and ionosopheric sounding also feature in the programme of research. In addition to R&D work, the Laboratories advise and guide the electronics industry in the incorporation of new designs in their products; established national standards on telecommunications materials, devices, and techniques; conduct quality control and testing work; and also advise the Ministry's agencies in the running of the telecommunications services. In training their staff, the Laboratories rely heavily on research organizations in the USA, such as the Bell Laboratories and the Western Electric Company.

The Ministry of Communications is also responsible for the Central Meteorological Bureau. The Bureau is entrusted with the provision of routine meteorological services. It also sponsors some research on ground levels and high altitude atmospheric conditions, typhoon precipitation, cloud physics, etc.

In 1970 a project on typhoons was set up in the Bureau jointly with the Academia Sinica (Institute of Physics), the National Taiwan University (Geography Department), and the Taiwan Air Force Weather Centre. The programme involves the collection and analysis of typhoon data; studies on their structure, formation, and dynamics; relevant experiments concerning sampling and modification of typhoons; typhoon forecasting; typhoon damage caused by winds, precipitation and tidal waves; as well as the study of atmospheric science in general. The existing meteorological network of radar stations, precipitation monitoring stations, data dissemination structures, and telecommunications systems are used wherever possible.

Ministry of Economic Affairs

The Bureau of Commodity Inspection and Quarantine provides for industry a comprehensive technical inspection service for materials and products. Laboratories exist for work on agricultural livestock (see section 6), food, marine products, chemistry, electrical and mechanical engineering, and materials testing. Research is performed on infectious diseases in animals, particularly serological diagnosis, and investigation of parasitic nematodes. Also under the Ministry of Economic Affairs is the important Industrial Technology Research Institute, to be described in Chapter 9.7.

National Bureau of Standards

The Bureau, in addition to administering the national standards, weights and measures, patents, and trademarks, conducts research into measurement of the standards themselves.

Central Geological Survey

Formerly named Geological Survey of Taiwan Province, this agency is responsible for all geological mapping, investigation of mineral and ground-water deposits, and research into general mineralogical, palaeontological, and petrographical fields.

Earthquake Research Project

Taiwan is situated in the Pacific Basin earthquake belt, and has long monitored earthquakes. The Earthquake Research Project was set up in 1970 in order to have a full-time establishment devoted to earthquake monitoring and earthquake engineering. The Central Meteorological Bureau runs a seismological network of 17 stations, with headquarters at Taibei and linked, since 1982, by a computerized telemetry system purchased from the USA.

Taiwan Provincial Government Bureau of Communications

The Railway Administration, the Highway Bureau, and the Harbour Bureau come under this administration. Research studies concern railway electrification, road bridge design, littoral sediment transport in harbour construction, and general basin modelling.

Academia Sinica

The institutes of the Academy cooperate with various universities in their research activities, and jointly maintain various research centres for the National Science Council. The science research institutes of the Academy are given below, with the scope of their research activities.

An institute is being set up for medicine: see section 5.

Institute of Botany: the Institute has various experimental farms and a herbarium, and performs studies in general biochemistry, biometry, cytogenetics, microbiology, virology and the breeding, nematology, pathology, physiology and tissue culture of plants.

Institute of Zoology: among the wide-ranging research of the Institute may be mentioned the study of chromatin and gene regulating molecules, fish enzymology, hormonal control of the reproductive physiology of rodents, reproductive physiology of cows, ecology of atomic power station effluent areas and intertidal pools, fish nutrition, toxicity of puffer fish, cytological effects of gamma rays on the oriental fruit fly spermatogenesis, morphology speciation, and isolation of drosophilia flies.

Institute of Biological Chemistry: allied to the Institute of Biochemical Science at the National Taiwan University, this newly established institute performs research in the areas of agricultural chemistry and pharmacology, eg, banana sucrose synthetase reaction and studies on phospholipase A.

Institute of Chemistry: both pure and applied research is conducted at the Institute, the latter emphasizing drug research and food research. In addition to organic, inorganic and microbial chemistry, there is considerable interest in marine chemistry.

Institute of Earth Sciences: formally established in 1982. Its current work is oriented towards the regional study of geology for the purpose of locating groundwater and fossil fuel resources. It undertakes research on earthquake forecasting in collaboration with the University of California at Berkeley and the University of Southern California. Taiwan seismologists could have learned much from their colleagues on the mainland if given the opportunity, however.

Institute of Physics: fluid dynamics, seismology, solid-state physics and medium-energy physics are the principal areas of investigation. The subfields being investigated are:

1 Typhoon atmospheric motions; ship hydrodynamics; pollution diffusion phenomena

2 Geological and seismic mapping; crustal deformations; earthquake forecasting through premonitory ground movements

3 Radiation-induced defects in semiconductors; solid-state spectroscopy; impurity levels in semiconductors; dielectric properties of biological membranes and acupuncture physics

4 Elastic scattering of protons; deuteron-induced reactions; spectroscopy of nuclei of light and medium masses.

Institute of Information: the latest to be founded, in 1983, its staff already have published reports on the subjects of Chinese character recognition and computer multilink terminals. Future work is directed towards automation theory, computer graphics, software reliability, operating system design, data base management and automation in the office.

Institute of Mathematics: there are three divisions in the Institute, namely, the Computer Science Division, the Mathematical Statistics Division, and the Pure Mathematics Division. The fields of probability and statistics figure prominently in the research work of the Institute; a group is also investigating computer input/output and software using Chinese characters.

9.4 Academic Research

There is research performed both in the departments of universities and technical institutes, and also in special research institutes attached to the larger universities. The Ministry of Education has designated 14 'key' institutions of higher education. Among them are four comprehensive universities, namely Taiwan, Chengkung, Zhongseng and Zhongshan Universities, and six specialist colleges: Yangming Medical, Oceanographical, Technological, National Normal, Taibei Normal and Gaoxiong Normal. The rest offer a range of courses but for research each has a single subject of emphasis, thus Qinghua University for nuclear technology, Polytechnic University for electronics, Central University for atmospheric science, and Political University for social sciences. At some of the establishments which are listed below, the accent is on teaching rather than research, especially at private institutions ie, those without 'National' in their full titles.

National Taiwan University at Taibei is the largest, with nearly 15 000 students and 2 000 instructors. Its predecessor was the Taihoku Imperial University founded in 1928 during the Japanese occupation. It has the following science-related colleges (departments are given in parentheses):
College of Medicine (anaesthesiology, anatomy, bacteriology, biochemistry, clinical pathology, dentistry, dermatology, gynaecology, internal medicine,

medical technology, neuro-psychiatry, nursing, ophthalmology, oto-rhino-laryngology, paediatrics, parasitology, pathology, pharmacology, pharmacy, physiology, public health, radiology, rehabilitation medicine, surgery, and urology). A teaching hospital is attached.

College of Agriculture (agricultural chemistry, agricultural economics, agricultural engineering, agricultural extension, agronomy, animal husbandry, forestry, horticulture, plant pathology and entomology, veterinary medicine).

College of Science (atmospheric sciences, botany, chemistry, geography, geology, mathematics, physics, psychology, and zoology).

College of Engineering (chemical, civil, electrical and mechanical engineering).

There are graduate schools in all of the foregoing disciplines, together with the following additional topics: biochemical science, microbiology, naval architecture engineering, oceanography, and pharmaceutical sciences. An affiliated institute of oceanography was established in 1968, and now comprises three divisions, devoted to physical oceanography, marine geology and geophysics, and biological oceanography and fishery, respectively. The latest addition, in 1983, is an institute for atmospheric science.

National Qinghua University at Xinzhu (where the science park is) concentrates on nuclear technology. Its enrolment is around 2 000 and its teaching staff numbers just under 300. It has its own reactor, with some equipment from the International Atomic Energy Agency. In addition to its undergraduate teaching, there are affiliated research centres devoted to chemistry, mathematics, nuclear engineering, nuclear science, and physics.

National Chengkung University at Tainan, the old provincial capital, is a technological university with an enrolment of over 10 000 and a faculty of 750 plus. The College of Engineering, with 4 000 students, is the largest of the five constituting colleges. Ten research institutes are affiliated, for the respective fields of: engineering science, architectural engineering, chemical engineering, civil engineering, electrical engineering, hydraulic and ocean engineering, industrial management, mining, metallurgy and materials science, mechanical engineering, and physics.

Fujen Catholic University at Taibei has graduate schools in mathematics and physics in its broad-based curriculum. Probably the richest of all the private universities, Fujen takes in about 12 000 students and employs over 1 000 academics.

Donghai Christian University, Taizhong, has a science and an engineering college, with departments of biology, chemistry, physics, architecture, architectural engineering, chemical engineering, and industrial engineering. A staff of about 1 400, some of whom work part-time (lecturing also at other universities), give full-time as well as evening courses to a total of over 9 000 students.

National Zhongseng University, also at Taizhong, was formed by the amalgamation of the Taiwan Provincial Colleges of Agriculture, and Law and Commerce. Its student body now approaches 10 000 and the number of teachers, 1 400. It has two science-related colleges, namely Agriculture, and Science and Engineering. Departments within the former cover agricultural economics, agricultural education, agronomy, animal husbandry, entomology, food technology, forestry, horticulture, plant pathology, soil and fertilizers, and soil and water conservation. Departments in the College of Science and Engineering cover applied mathematics, botany, chemistry, civil engineering, and mechanical engineering. One of the research units affiliated with the University is devoted to agricultural economics.

National Polytechnic University, Xinzhu, has a college of engineering which caters for marine technology, civil engineering, mechanical engineering, telecommunications, control engineering and electronics. The College of Science has departments in electrophysics, computer science and applied mathematics. There is a graduate school of traffic and transportation. The Institute of Electronics performs research in lasers and semiconductors, being one of the centres of excellence designated by the government for expansion. There is also an Institute of Management Sciences. The Institute of Applied Chemistry was newly founded in 1982. Nearly 300 of the university staff give classes, attended by 2 300 students.

The Zhongyuan Institute of Science and Engineering is located at Zhongli, a town midway between Taibei and Xinzhu. It has nearly 3 000 students and over 200 instructors. Since 1982 a research centre for hydrology and one for civil engineering have been set up.

National Central University, also at Zhongli, has departments in atmospheric physics, chemical engineering, civil engineering, mathematics, and physics. To it are attached the Institutes of Atmospheric Physics, Geophysics, Dataelectronics, and Mechanical Engineering, which are in effect graduated schools. The Department and Institute of Atmospheric Physics are due for a major expansion. Students number just over 1 100 and teachers, about 170.

Dongwu University, or Suzhou University as it is differently called in English, is in the city of Taibei. Its College of Sciences accommodates some 1 500 students, taught by about 80 instructors, a large fraction of whom hold part-time appointments (teaching in more than one universities). The other colleges are for arts, law, commerce and Chinese literature; total enrolment is nearly 10 000.

National Zhongshan University at Gaoxiong is the smallest of the designated key institutions of higher education. There are under 1 000 students on the campus.

Fengchia College of Engineering and Business at Taizhong has the science-related departments of applied chemistry, architecture, computer science, statistics, and mathematics. Engineering departments are devoted to civil, electronic, hydraulic, mechanical and textiles engineering. Undergraduates number 14 000 and research students nearly 100. In 1980 an independent department was created for environmental science, the only one so far found in a Taiwan university. The subject for specialization in the final year of the course is environmental protection, ecology or environmental engineering.

National College of Technology at Taibei comprises the departments of chemical, textile, mechanical, electrical and electronic engineering and the department of industrial management. There are nearly 2 000 students and 100 faculty members.

National Normal University, also at Taibei, has about 1 000 part- and full-time lecturers teaching 9 000 day and evening students.

There are Provincial Colleges of Agriculture (at Taizhong), Oceanography (at Jilong near Taibei), and Nursing (at Taibei). There is a Provincial Institute of Agriculture at Pingdong near Gaoxiong.

9.5 Medicine

Public health and medical research goes on at major institutions of higher education such as National Taiwan University and Yang Ming Medical College, as well as at military organizations like the Army Veterans General Hospital and the National Defence Medical Centre; all these establishments are in Taibei. The medical school in Taiwan University has already been described in the last section. Yang Ming Medical College dates from 1975 and has about 800 students taking its seven-year courses, the clinical part of which is held at the Army Veterans General Hospital. This hospital contains a research division named Cooper, after a certain American missionary who was once posted to Taiwan. The National Defence Medical Centre instructs in dentistry, surgery, internal medicine, pharmacy and nursing.

Besides the above national establishments, there are private medical colleges at Taibei and Gaoxiong. The China Medical and Pharmaceutical College at Taizhong has departments of Chinese medicine, medicine, nursing, and pharmacy. Students are counted as about 2 500 and instructors, full- and part-time, over 250.

Results that have been published by Taiwan researchers over the last few years mostly relate to local problems such as sanitation, nutrition including food enrichment and metabolic consequences, as well as local diseases like filariasis and endemic arsenic poisoning. Studies in contraception and abortion have also been reported: the government is undertaking a 'mild' birth control programme. There has also been work on snake venom, involving the production of anticoagulants from snake toxins and the isolation of protein from cobra toxin.

Fundamental research is being conducted on cancer. A group with members from Yang Ming Medical College and the Veteran General Hospital claimed in 1983 to have beaten the world in preparing the first monoclonal antibody for liver cancer. Much attention is given to the antitumour properties of Taiwan flora, particularly those plants certain parts of which have been used in traditional Chinese medicine. In this work collaborations have been sought with the US National Institute for Health. The Academia Sinica is setting up a new Institute of Medical Sciences at Taibei. One of the three emphases of the work will be on the molecular biology approach to cancer research; the others are to be coronary diseases and hepatitis.

9.6 Agriculture

In the so-called Twelve Major Development Projects (see next section) being undertaken by the Taiwan government, three are related to agriculture. They are: agricultural mechanization, expansion of the drainage system, and construction of dykes on the west coast and inland levees. For research, organizations which are active in the fields of agricultural science and technology are described below. Establishments that have already been mentioned in the sections on government and academic research will be omitted in the following, however.

Animal Industry Research Institute: the Ministry of Economic Affairs runs this establishment, whose activities cover breeding, disease, livestock management, nutrition, and use of waste, not only from the research and the technical assistance aspects but also from the point of view of technical training. Among research projects there are: inbreeding studies on swine crossing and use of frozen semen; nutrition studies on animal diets, effects of protein levels and amino acid levels in productivity of sows, and use of converted swine waste as a source of nutrients; veterinary disease studies on acupuncture in pigs, aflatoxin, haemorrhagicnecrotic enteritis in swine and pseudorabiles control in hogs.

Food Research Institute: the objectives of this Institute are to encourage expansion of the indigenous food industry by feasibility and productivity surveys, and quality control investigations of new processed foods. Food processing has been listed as a priority topic of research in the 1982–85 Science Plan.

Taiwan Agricultural Research Institute: the Taiwan provincial government administers this major establishment, which dates from 1895 and which now has a research staff of about 250. There are six operational departments which cover agronomy, horticulture, agricultural chemistry, plant pathology, applied zoology, and agricultural engineering. The main activities encompass research on crop improvement, including breeding and cultural methods; investigation of plant physiology, genetics, and cytogenetics; soil surveys, soil microbe and plant nutrition research; soil fertility, and fertilizer analyses; the chemistry of agricultural products, and nutriophysiological problems of rice plants; pest and plant disease research; and modernization of agricultural machinery. There are three specialized centres performing experimental work: Chiayi Agricultural Experiment Station, which conducts research mainly on rice, sweet potatoes and fruit trees; Tainan Fibre Crops Experiment Station, performing research into fibre crops breeding and cultivation, their protection and processing; and Fengshan Tropical Horticultural Experimental Station, carrying out research on pineapple and vegetable breeding, cultivation, and processing. Incidentally, the Institute of Nuclear Science in the Qinghua (Taiwan) University, to be described in Chapter 9.8, carries out some mutation breeding work.

Taiwan Forestry Research Institute: also founded in 1895 as a nursery, this organization employs about 150 researchers and has eight research divisions investigating the biology, chemistry, economics, growth, management, and utilization of forests and forest products. The aims of the work are the improvement of management of natural hardwood and coniferous forest to achieve their optimum utilization; the acquiring of information on

commercial indigenous and exotic tree species for future planting programmes; the systematic study of seed trees, seeds, seedling growing,
planting, tending, logging, and utilization; the study of forest diseases and
pests; forestry economics; and the physical, mechanical and chemical
properties of forest products, with emphasis on wood utilization, preservation and seasoning. The headquarters of the Institute are at Taipei, along
with six outstations.

Taiwan Fisheries Research Institute: founded in 1909, it has five
outstations in addition to its fisheries laboratory. There are around 150
technical personnel. General research is performed in support of the fishing
industry of Taiwan. For example, the region's pelagic fishing grounds are
explored, and improvements are made in fishing technology methods. In
addition to research into fishery resources (local grey mullet, mackerel and
tuna) there are extensive fish culture studies, Chinese carp, eels, mullet,
oyster and trout.

Taiwan Sugar Research Institute: located in Tainan, the Institute is an
example of research centres supported by public corporations. In addition to
studying sugarcane genetic breeding, plant nutrition and plant protection,
and raising the production efficiency of sugar mills, it carries out R&D into
antibiotic and enzyme reactions, the preparation of ribonucleic acid from
yeast, and the use in industry of cane sugar byproducts, such as molasses
and bagasse. It has three experimental outstations, at Huwei, Sengying and
Pinding.

9.7   Industrial Research

Current industrial expansion is highlighted by the Twelve Major Development Projects that were announced by the government in 1977. Of them
one, in heavy industry, is the modernization of the China Steel Mill at
Gaoxiong: it is the only project completed so far. The three relating to
agriculture have been itemized in the last section. Five others concern
communications, namely the enlargement of Taizhong harbour, completion
of the round-the-island railway, construction of a cross-island motorway,
widening of the Pingdong-Dabanlie road link and an improvement scheme
for the Kapping regional traffic. Another infrastructural development is the
installation of two nuclear power stations near Taibei and Dabanlie
respectively. The remaining projects refer to the creation of several satellite
towns around Taibei, and an urban housing programme.

In research, most of the feasibility studies, technology transfer, innovation sustenance and product development are carried out in a series of
governmental establishments, which come under the Ministry of Economic
Affairs, with back-up from the National Science Council. Principal among
the organizations is the Industrial Technology Research Institute (ITRI),
which was formed in 1973 by amalgamation of three Ministry institutes, viz,
Metals Industry Research Laboratories, Mining Research and Services
Organization, and Union Industrial Research Laboratories. The purpose of
the grouping was to establish a new autonomous organization, better

equipped and more flexible, to deal with the promotion of technology in Taiwan. Owing to the gap of an electrical side in the merger, the Electronics Industry Research Centre was added in 1974. In 1981 the Energy Resources Research Centre became the fifth division. The latest addition is the Industrial Materials Research Centre, set up in July 1983. It is devoted to basic investigations (fatigue, fracture mechanics, etc), non-destructive testing of solids, and the processing technology of special steels and alloys for armaments refractory ceramics, electronic materials and high polymers. The first on the list of its current projects is the fabrication of graphite fibre composites for use on supersonic fighter aircrafts and tactical missiles. Technical staff in the entire Institute has increased from 400 in 1973 to the present number in excess of 2 400.

The objectives of ITRI are:
1 to perform industrial research work and technical services as requested by the armed forces,
2 to conduct R&D into new industrial products and production methods and ensure their practical application,
3 to investigate and analyse industrial development and market situations,
4 to coordinate with bodies concerned with industrial R&D,
5 to investigate, collect, and publicize new industrial technology from abroad,
6 to conduct mineral exploration and technical assistance for the mineral industry,
7 to review and conduct feasibility studies on new industrial technology, and economic evaluation of industrial projects, and
8 to conduct industrial R&D.
As with the Academia Sinica, ITRI has close liaison with other research institutions and universities. There are collaborations in fields such as agriculture and telecommunications. Such work, along with the main body of contract research investigation, is funded by industry, government, and private firms. ITRI performs services to existing industries as required, and develops new products and processes to a pilot plant stage. If these projects have techno-economic potential they are offered to private firms or investors for full-scale industrialization. ITRI may be requested by the government to undertake larger-scale production activities where the manufacturing technology is new to Taiwan, extensive special training is necessary, or the economic reward is long-term. When the yields and costs of such a venture indicate a potentially economically viable concern, then it is offered to entrepreneurs for commercial development.

The bias of skills towards minerals, mining, metallurgy, and engineering, which ITRI inherited from its three constituent bodies has been balanced by the activities of the three newer research centres. Some of the on-going projects are: improvement of local clays, heavy sands, and iron pyrites for use as insulators, production of rare earth metals, manufacture of manganese-zinc and nickel-zinc ferrite bodies for use as television components, processing of oxide ceramics, manmade fibre fabrication, integrated circuits design, development of micro- and minicomputers, geothermal energy, and feedstuff production from waste materials. A pilot plant which produces magnesium hydroxide from seawater concentrate has been designed.

Other nationalized R&D establishments include the Glass Products Research Institute, set up jointly in 1966 by the Council for International Economic Cooperation and Development and the Xinzhu Glass Works. This establishment serves the glass industry across a whole range of testing and analysis services, quality control, and development work on optical glass, high-temperature glass, and glass fibre technology. The Taiwan Alkali Company is interested in the development of chlorinated chemical products and the Taiwan Aluminium Corporation is concerned with ore extraction and reduction. The Taiwan Power Company ('Taipower') Research Institute conducts general research in the field of heavy-current electrical engineering.

Among private industries the following firms carry out R&D: the Nan Ya Plastics Corporation, the Taiwan Cement Corporation, the Taiwan Pulp and Paper Corporation, and the Wei Chan Food Corporation. Other firms make regular financial contributions to government research centres, eg, the Taiwan Fertilizer Company supplies some funds for the Taiwan Agricultural Research Institute described in the last section.

9.8   Energy

The Ministry of Economic Affairs has announced its aim to reduce Taiwan's reliance on petroleum. Correspondingly, it wants the coal and nuclear sectors to increase their contribution. Coal reserves are found mostly in the north and central areas of the island. However, Taiwan coal in general is of young geological age and low degree of carbonization, especially the varieties from the central mines, which typically have low caloricity but high ash and sulphur contents. Furthermore, after 1964–68 during which period domestic coal production exceeded 5 megatonnes a year, equivalent to around 60 per cent of the annual energy consumption, the annual output decreased continuously for nearly a decade. This downward trend was reversed only in 1976. At present, the output remains under 3 megatonnes per year and provides for less than 40 per cent of primary energy. In accordance to the policy of shifting reliance from hydrocarbon, all of the new fossil-fuel stations being built by Taipower are coal-fired. The additional demand for coal will have to be satisfied by imports, at least in the near future. The government is keen to secure foreign sources of supplies. The Mining Research and Services Organization (MRSO) in ITRI, described in the last section, has participated in coal exploration in Indonesia.

The other joint venture undertaken by MRSO is uranium prospecting in Paraguay. Taiwan is not known to have uranium deposits nor to operate nuclear fuel reprocessing plants. Its first thermonuclear power station was constructed in 1970. The Ministry of Economic Affairs plans that, of the total electricity-generating capacity of some 20 gigawatts on the island, a staggering fraction of 40 per cent will be provided by eight generators distributed through three nuclear stations. As said before, the erection of stations number 2 and 3 is in an advanced stage. All the eight generators are powered by reactors of the pressurized water type; for the future second-generation reactors (breeder reactors) have been talked about. To make the enormous investment needed by the nuclear porgramme Taipower has borrowed heavily, chiefly from US banks and finance houses. In return, American corporations have dominated the contracts for all the eight reactor-turbogenerator assemblies.

The Institute of Nuclear Energy Research is the R&D arm of the Atomic Energy Council, which as explained earlier reports directly to the Office of the Prime Minister. The practical research activities of the Institute are applied to reactor technology, power plants, and fuel cycle recovery problems. A 40-megawatt heavy-water reactor and a 7.7-megaelectronvolt van de Graaf generator are available for research. Other branches of the Council include the Planning Division, which is concerned with nuclear legislation, national and international cooperation, training, and information collection and dissemination. The Technical Division performs the supervisory and regulatory activities, being responsible for all radiation protection and inspection, safety evaluation, licensing and supervision of nuclear facilities, and building and operation. It is also in charge of promoting the practical application and development of nuclear energy.

The Atomic Energy Council dates from 1955 and in the following year the Institute of Nuclear Science was established within the graduate school of the Qinghua (Taiwan) University. The Institute provides irradiation services via a 1-megawatt swimming-pool research reactor and supplies radioisotopes. Teaching is done in conjunction with the faculties of the University. General research fields include radiation chemistry, neutron activation analysis, and radioactive fall-out. Improved rice varieties have been produced by gamma-radiation induced mutation. Such irradiation studies fall also within the interests of the Union Industrial Research Laboratories of ITRI, there being applied to preservation of foodstuffs, sterilization of pharmaceutical products, and in other areas.

Geothermicity and biogas are the two forms of alternative energy that have been exploited in Taiwan. Geothermal sites have been located in places like Hualian on the eastern coast. The MRSO of ITRI has been involved in their development, and designed power plants of capacities up to 1 megawatt. ITRI's newer Energy Resources Research Centre confines its interests to solar, wind, seawave and biomass energy sources. Only the last form has been exploited on the practical level. On the island there are now several thousand biogas digesters; much can be learned from mainland China, which is an acknowledged world leader regarding experience with biogas utilization.

9.9 Defence

In this category, the most important organization for basic research is the Zhongshan Academy of Sciences. Established in July 1969, it occupied a secluded corner of Shimen Reservoir which is 40 kilometres west of Xinzhu. Topics covered in its work include explosives chemistry and physics, shock mechanics, propellants and propulsion, metallurgy, and electronics. Among the completed projects itemized in its 1981 Annual Report are those connected with the 'Kunwu' antitank guided missile, infrared scanners and precision casting techniques. The Kunwu, the first ATGM to be indigenously produced, has laser guidance and a range of over 3 kilometres.

The Aero Industry Developing Centre (AIDC) at Taizhong is the main R&D centre for military aircrafts. Set up also in 1969, it was originally administrated by the Air Force but, since 1982, has become subordinate to the Zhongshan Academy just described. Like other establishments of military research, it has over the past several years been endowed with increasing funding. One of its major research projects was the development of C-2, a light military transport using twin AVCO Lycoming T-53-L-701A turboprops, which it makes locally under license. The project was unsuccessful and has now been terminated. There were recently some preliminary negotiations concerning the purchase of the C-160 Transall, a Franco-Germanic production, but the move appears to have been stopped because of political factors. The backbone of the ground-attack fighter wing of the Taiwan Air Force consists of F5E's that are domestically manufactured under licensing by the US. The F5E carries two 20-millimetre M39/A2 cannons, and either two Sidewinder or Shafrir AAM, or two Bullpup or Maverick ASM. A daylight Mach-1.6 fighter, it is, as claimed by the Taiwan authority, inferior in combat performance to the J-8 and even the J-7 of the People's Republic. Late in 1981, after the Regan administration finally decided against selling the much more advanced F5G to Taiwan so as to comply with the Shanghai II Communique, Northrop Inc was asked to assist AIDC in upgrading the F5E the latter has been coproducing. The modifications being implemented by AIDC include reshaping the radome in imitation of the F5G's 'Tigershark' nose, and aerodynamic improvements to increase the lift, thus raising the maximum climb rate to 165 metres per second at sea level. Limited all-weather operational capacity is achieved by the incorporation of certain electronic equipments, and the replacement of the original APQ-153 radar by the planar APQ-159 radar in the radome.

The Army Institute of Technology is the technical training branch of the army. Some engineering research work is performed, such as the use of satellites for communications purposes. The Naval Institute of Technology performs a similar role for the naval arm, teaching naval architecture and marine and electrical engineering. It is located at Jilong which, together with Qoying, and Magong on one of the Penghu islands, are the three chief naval bases. Research interests are in the fields of navigation aids, stability of control systems, and maritime communications.

Appendix 1

Statistical Tables

Table 1 Human and natural resources (1981)

Source: State Statistical Bureau figures as reported in 10 January 1983 issue of *Beijing Review*.
Note: In the table 1 billion = 1 000 millions.

1. **Population**
   Total population at the end of the year   1 014.36 million
   Population of national minorities   62.18 million
   Density   104   per sq. km.

2. **Land**
   Area of territory   9.6   million sq. km.
   Topography distribution
      1) Mountains   33   per cent
      2) Plateaux   26   per cent
      3) Basins   19   per cent
      4) Plains   12   per cent
      5) Hills   10   per cent

3. **Climate**
   Annual average rainfall   630   mm.
   Total annual rainfall   6 000   billion cu. m.
   Climate distribution
      1) Wet area (with drought index less than 1.0)   32   per cent
      2) Semi-wet area (with drought index between 1.0–1.5)   15   per cent
      3) Semi-dry area (with drought index between 1.5–2.0)   22   per cent
      4) Dry area (with drought index more than 2.0)   31   per cent

4. **Forests**
   Total area of forests   119.78 million hectares
   Nation's area covered by forests   12.5 per cent
   Total deposits of timber   9.35 billion cu. m.

5. **Grasslands**
   Total area of grassland   319.08 million hectares
   Usable area   224.34 million hectares

Table 1 – *continued*

6. **Hydrology and water conservancy**
   1) Total annual runoff of rivers: 2 614.4 billion cu. m.
      Zhujiang (Pearl) River basin   307   billion cu. m.
      Changjiang (Yangtze) River basin   979.3 billion cu. m.
      Huaihe River basin   53   billion cu. m.
      Huanghe (Yellow) River basin   56   billion cu. m.
      Haihe River basin   28.4   billion cu. m.
      Songhua River basin   75.9   billion cu. m.
      Rivers in Zhejiang and Fujian Provinces   200.1   billion cu. m.
      Rivers in Tibet   359   billion cu. m.
   2) Total area of fresh water   16.64 million hectares
      Area suitable for breeding aquatic products   5.03 million hectares
      Area used for breeding aquatic products   2.74 million hectares
   3) Water resources reserves   676   million kw
   4) Area of sea fishing ground   818 000 hectares
   5) Sea area for breeding marine products   492 000 hectares
      Area used for breeding marine products   116 000 hectares
   6) Mainland coastline   18 000   km

7. **Mineral resources**
   Coal deposits   642.7 billion tons
   Iron ore deposits   44.31 billion tons
   Number of minerals: 134
      Energy (coal, petroleum, gas, etc)   6
      Ferrous metals   5
      Nonferrous metals   20
      Rare or rare-earth metals   28
      Nonmetal minerals   75

Table 2   Administrative divisions and population distribution

Source: State Statistical Bureau figures as reported in *Beijing Review*, 3 January 1983.
Notes: The population given for Fujian (13) includes those of Jinmen, Mazu and adjacent islands, which add up to 57 847; the population for Guangdong (19) excludes those of Dongsha and Nansha Islands, which have not been counted in the census, and those of Xianggang (Hong Kong) and Aomen (Macao), which add up to 5 378 627; not included in the population figures is the number in the armed services, which total 4 238 210.

No.	Units at the Provincial Level	Capital	Area (1 000 sq. km.)	Population	Units at the Prefectural Level	Cities	Units at the County Level	Districts Under the Cities
	Total: 30		**9 600**	**1 031 882 511**	**208**	**230**	**2 136**	**514**
1	Beijing		16.8	9 230 687			9	10
2	Tianjin		11.3	7 764 141			5	13
3	Hebei Prov.	Shijia-zhuang	180+	53 005 875	10	10	139	39

Table 2 – *continued*

No.	Units at the Provincial Level	Capital	Area (1 000 sq. km.)	Population	Units at the Prefectural Level	Cities	Units at the County Level	Districts Under the Cities
	Total: 30		**9 600**	**1 031 882 511**	**208**	**230**	**2 136**	**514**
4	Shanxi Prov.	Taiyuan	156	25 291 389	7	7	101	15
5	Inner Mongolian Autonomous Region	Hohhot	1 200	19 274 279	9	10	79	13
6	Liaoning Prov.	Shenyang	140+	35 721 693	2	13	45	42
7	Jilin Prov.	Changchun	180+	22 560 053	4	9	37	9
8	Heilongjiang Prov.	Harbin	460+	32 665 546	7	12	66	61
9	Shanghai		6.2	11 859 748			10	12
10	Jiangsu Prov.	Nanjing	100+	60 521 114	7	11	64	33
11	Zhejiang Prov.	Hangzhou	100+	38 884 603	7	9	62	13
12	Anhui Prov.	Hefei	130+	49 665 724	8	12	70	34
13	Fujian Prov.	Fuzhou	120+	25 931 106	7	7	61	10
14	Jiangxi Prov.	Nanchang	160+	33 184 827	6	10	81	16
15	Shandong Prov.	Jinan	150+	74 419 054	9	9	106	21
16	Henan Prov.	Zhengzhou	167	74 422 739	10	17	111	34
17	Hubei Prov.	Wuhan	180+	47 804 150	8	11	73	13
18	Hunan Prov.	Changsha	210	54 008 851	11	14	89	22
19	Guangdong Prov.	Guangzhou	210+	59 299 220	9	14	96	18
20	Guangxi Zhuang Autonomous Region	Nanning	230+	36 420 960	8	7	80	17
21	Sichuan Prov.	Chengdu	560+	99 713 310	14	13	182	22
22	Guizhou Prov.	Guiyang	170+	28 552 997	7	5	79	5
23	Yunnan Prov.	Kunming	390+	32 553 817	15	6	123	4
24	Xizang Autonomous Region	Lhasa	1 200+	1 892 393	5	1	71	1
25	Shaanxi Prov.	Xian	200	28 904 423	7	6	91	11
26	Gansu Prov.	Lanzhou	450	19 569 261	10	5	73	6
27	Qinghai Prov.	Xining	720+	3 895 706	7	2	37	4
28	Ningxia Hui Autonomous Region	Yinchuan	60+	3 895 578	2	2	16	7
29	Xinjiang Uygur Autonomous Region	Urumqi	1 600+	13 081 681	12	8	80	9
30	Taiwan Prov.	Taibei	36+	18 270 749				

Table 3   Agricultural productions in perspective

Source: Compiled from figures quoted in official speeches or released by the State Statistical Bureau

Year	Gross National Output Values Industrial*	Agricultural*	Grain**	Oil-Bearing Crops**	Cotton**	National Population
1957	70.4	53.7	195	4.2	1.64	647 m
1965	139.4	59.0	200	?	1.7	726 m
1978	455	81	305	5.22	2.17	958 m
1982	630	120	340	12.2	3.3	1 032 m

* in thousand million 1952-constant yuan   ** in million tonnes

Table 4   Primary energy reserves estimates

COAL

Reserves (in 10^9 tonnes)	Category	Source
643	verified	State Statistical Bureau, 1982
>590	proved	Ministry of Coal Industry, at Int Mine Planning and Development Symp, Beijing, Sept 1980
1425	potential	World Energy Conf, 1978

CRUDE OIL*

Reserves (in 10^9 tonnes)		Category	Source
5.6	3.8 onshore / 1.8 offshore	potential	M. J. Terman, in SEAPEX Proc., 1976, p 127
7.8	3.7 onshore / 4.1 offshore	potential	A. A. Meyerhoff and J.-O. Willums, CCOP Tech. Bull., 10, 1976
10.6	5.3 onshore / 5.3 offshore	proved	J. Emerson, Chase Manhattan Bank energy economist, 1982

NATURAL GAS*

Reserves (in 10^{12} m^3)		Category	Source
0.7		proved recoverable	Oil & Gas J., 29 Dec 1980
8.76	5.66 onshore / 2.9 offshore	ultimately recoverable	Meyerhoff and Willums, op. cit.; Petroleum Economist, Nov 1981, p 484

HYDRO-ENERGY

Reserves (in 10^9 MWh)	Category	Source
5.92	reserves	State Statistical Bureau, 1982
2.63	industrially exploitable under conditions of average flow	Y. I. Berezina, 'Toplivno-Energeticheskdya baza Kitaiskoi Narodnoi Republiki', IVL, Moscow 1959, p 58

* Figures are those given for year 1983 start, and where necessary have been corrected for quantity extracted since original date of validity

Table 5 Recoverable reserves and output in 1982

	Reserves	(World share	World rank	S/CT)*	Production	(World share	World rank	S/CT)*
COAL	643 Gt	30%	2	98%	0.62 Gt	20%	2	72%
OIL	2.6 Gt	3%	9	0.8%	0.1 Gt	3%	10	22%
GAS	$0.7 \times 10^{12} m^3$	10%	6	1%	$1.3 \times 10^{10} m^3$	3%	6	3%
HYDRO	$0.59 \times 10^{10} MWh$	30%	1	0.2%	$0.48 \times 10^8 MWh$	2%	9	3%

* Share of China's total, calculated with the use of the equivalence relationships:
1 Gt (10^9 tonnes) of coal = 0.67 Gt oil = $0.78 \times 10^{12} m^3$ gas = $0.8 \times 10^{10} MWh$

Table 6 Crude coal production

Sources: Compiled and calculated from data in 'Energy Balance Projection for the PRC 1975–1985', National Council for US–China Trade, Washington DC 1975, and 'China: Economic Indicators', CIA, Washington DC 1978; and where available official figures from State Statistical Bureau, 1981 and 1982. Note: * According to K. Woodard, 'The International Energy Relations of China', Stanford University Press, Stanford 1980, p 536

	Output (in Mt)	S/CT	Contribution from small mines	Foreign exports (net, in Mt)*
1949	32.43	98 per cent	32 per cent	?
1957	130.7	96 per cent	6 per cent	0.7
1960	280.0	95 per cent	28 per cent	0.5
1965	232.2	85 per cent	15 per cent	1.0
1970	327.4	76 per cent	25 per cent	0.5
1972	376.5	74 per cent	26 per cent	0.3
1974	409.0	67 per cent	28 per cent	0.5
1976	483.0	68 per cent	33 per cent	0.3
1978	585.0	70 per cent	38 per cent	0.4
1980	620.0	72 per cent	40 per cent	?
1981	621.5	71 per cent	44 per cent	?
1982	651.0	?	?	?

Table 7 Crude oil production and consumption

Sources: * State Statistical Bureau, 1981 and 1982
** 'BP Statistical Review of the World Oil Industry 1980', British Petroleum 1981
*** 'World Energy Supplies' 1971–75; 1972–76; 1973–78; 1975–80 United Nations

	Output (in Mt)		share of	Domestic consumption (in Mt)	
	*	**	world total**	**	***
1949	0.12	?	?	?	?
1970	30.2	28.2	1.2 per cent	28.2	21.77
1972	44.5	42.1	1.6 per cent	43.1	37.53
1974	64.2	65.8	2.3 per cent	61.9	52.42
1976	87.0	83.6	2.8 per cent	76.9	66.0
1978	104.1	104.1	3.4 per cent	84.7	79.23
1980	105.9	105.8	3.5 per cent	88.0	
1981	101.2				
1982	101.8				

Appendix 2

First Five-Year Plan, 1953

Source

Extracted from *First Five-Year Plan for Development of the National Economy of the People's Republic of China in 1953–1957* (Foreign Languages Press, Beijing, 1955) pp 21–39.

The sum of 42 740 million yuan for investments in the five-year capital construction programme is distributed as follows:

Industrial departments, 24 850 million yuan, or 58.2 per cent of the total amount to be invested;

Agriculture, water conservation, and forestry departments, 3 260 million yuan, or 7.6 per cent;

Transport, postal, and telecommunications departments, 8 210 million yuan, or 19.2 per cent;

Trade, banking, and stockpiling departments, 1 280 million yuan, or 3 per cent;

Cultural, educational, and public health departments, 3 080 million yuan, or 7.2 per cent;

Development of urban public utilities, 1 600 million yuan, or 3.7 per cent;

Other items, 460 million yuan, or 1.1 per cent.

. . . .

In the sphere of industry, we list below figures showing the ultimate increases in annual production capacity of principal industrial items when all the above-norm and below-norm construction projects started in the First Five-Year Plan period are completed; and figures showing the increases in annual capacity by the end of the First Five-Year Plan period when part of them are completed:

Pig iron: ultimate increase in annual capacity, 5 750 000 tons; increase in annual capacity by the end of the five-year period, 2 300 000 tons

Steel: ultimate increase in annual capacity, 6 100 000 tons; increase in annual capacity by the end of the five-year period, 2 530 000 tons

Electric power: ultimate increase in annual capacity, 4 060 000 kilowatts; increase in annual capacity by the end of the five-year period, 2 050 000 kilowatts

Coal: ultimate increase in annual capacity, 93 100 000 tons; increase in annual capacity by the end of the five-year period, 53 850 000 tons

Metallurgical and mining machinery: ultimate increase in annual capacity, 190 000 tons; increase in annual capacity by the end of the five-year period, 70 000 tons

Power-generating equipment: ultimate increase in annual capacity, 800 000 kilowatts. All projects will be completed within the five-year period.

Trucks: ultimate annual capacity, 90 000 vehicles; annual capacity by the end of the five-year period, 30 000 vehicles

Tractors: ultimate annual capacity, 15 000, to be reached in 1959

Chemical fertilizers: ultimate increase in annual capacity, 910 000 tons; increase in annual capacity by the end of the five-year period, 280 000 tons

Cement: ultimate increase in annual capacity, 3 600 000 tons; increase in annual capacity by the end of the five-year period, 2 360 000 tons

Cotton spindles: ultimate increase, 1 890 000 spindles; portion to be put into operation in the five-year period, 1 650 000 spindles

Machine-made paper: ultimate increase in annual capacity, 186 000 tons; increase in annual capacity by the end of the five-year period 95 000 tons

Machine-processed sugar: ultimate increase in annual capacity, 560 000 tons; increase in annual capacity by the end of the five-year period, 428 000 tons

In the sphere of transport, more than 4 000 kilometres of new trunk railways and branch lines will be built in the five-year period. If to this is added the mileage of railways to be restored, reconstructed, or double-tracked, extended station spurs, and industrial and other special lines, the total length of the railway network will be increased by some 10 000 kilometres. Upward of 10 000 kilometres of highways will be built or rebuilt with capital provided by the Central People's Government in the five-year period, and more than 7 000 kilometres will be opened to traffic. Four hundred thousand tons deadweight of new steamships will be acquired in the five-year period.

In the sphere of agriculture and water conservation, 91 mechanized state farms and 194 tractor stations (both above-norm and below-norm) will be set up in the five-year period. During this period, thirteen big reservoirs will be built. In addition, dredging of waterways and repairing of dykes will involve 1 300 million cubic metres of earth and masonry work, and we will begin the engineering project to harness the Yellow River.

Buildings with a total floor space of about 150 million square metres will be constructed in the five-year period, including factory buildings, housing for factory and office workers, schools, and hospitals.

The industrial construction programme is the core of our First Five-Year Plan, and the construction of the 156 industrial projects to be built with Soviet aid is in turn the core of the industrial construction programme. Within the period of the First Five-Year Plan, work will have begun on 145 of these 156 projects, while survey and design work will have been carried out on the remaining 11 projects, which will go into construction in the period of the Second Five-Year Plan.

These industrial construction projects are large in scale and new in technique. Many of them are unprecedented in the history of Chinese industry. For example, in the eight-year period between 1953 and 1960, the integrated iron and steel works in Anshan, building on the basis of its original capacity, will complete, in the main, the construction or reconstruction of the following forty-eight major projects: three iron ore mines, eight ore-dressing and sintering plants, six automatic blast furnaces, three modern steel-making plants, sixteen rolling mills, ten batteries of coke ovens, and two heat-resistant material shops. The latest achievements of Soviet technology will be utilized to the fullest possible extent in the building or reconstruction of these plants, mines, and shops. When its reconstruction is completed, this integrated iron and steel works – the biggest of its kind in China – will increase its annual capacity to 2 500 000 tons of pig iron, 3 220 000 tons of steel and 2 480 000 tons of rolled steel. Its output of steel plates, sheets, tubes, and other rolled steel of various specifications will, on the whole, be able to meet the country's requirements for the manufacture of locomotives, steamers, motor vehicles, tractors, etc, during the period of the First Five-Year Plan and the early years of the Second Five-Year Plan. Its annual output of rails of different specifications will be sufficient to lay more than 3 000 kilometres of railways.

Simultaneously with the reconstruction of the integrated iron and steel works in Anshan, construction will go ahead on two new integrated iron and steel works in Wuhan and Paotow. Fifteen thermal power stations, each with a capacity of more than 50 000 kilowatts, are among the power plants to be built in the five-year period. After reconstruction, the Fengman Hydroelectric Power Station will have a capacity of more than 560 000 kilowatts. Completion of these projects will vastly increase the supply of electric power in various regions.

Coal-mining enterprises to be built during the five-year period include thirty-one with a projected annual capacity (counting the original capacity) of more than 1 million tons of coal each. Among these, which include those designated in China, the annual capacity of the five biggest mining enterprises will reach the following levels by 1957: mines under the Fushun Mining Administration, 9 300 000 tons; mines under the Fuhsin Mining Administration, 8 450 000 tons; mines under the Kailan Mining Administration, 9 680 000 tons; mines under the Tatung Mining Administration, 6 450 000 tons; and mines under the Huainan Mining Administration, 6 850 000 tons.

The First Motor Works will be completed in the present five-year period. When it reaches projected capacity, it will be able to provide transport with 30 000 trucks a year. The Second Motor Works, with double the capacity of the first, will also begin construction within the period of the First Five-Year Plan. These two plants will lay the foundation for China's automotive industry.

When the tractor plant, which will go into construction in the present five-year period, is completed in the period of the Second Five-Year Plan, China will be able to produce annually 15 000 54-hp. tractors to meet the needs of agriculture.

When the two heavy machinery plants (one designed for us by the Soviet Union and the other by ourselves) that begin construction in the present five-year period are completed, they will be able, according to their projected capacities, to produce every year a complete set of iron-smelting, steel-making, rolling mill, and coke oven equipment for an integrated iron and steel works with an annual capacity of 1 600 000 tons of steel.

When . . . the plants making power-generating equipment to be started in the five-year period are completed, China will be able to manufacture 12 000-, 25 000-, and even 50 000-kilowatt power-generating units to meet the requirements of electric power development in all branches of the national economy.

Many of our light industry plants were designed and built by ourselves, and many of these are of considerable size. The three cotton mills that have been or will be built in our capital, Peking, for instance, will be equipped with 230 000 spindles and more than 7 000 looms. In all, thirty-nine textile mills of considerable size will be built in the five-year period.

Many of these new industrial construction projects are large in scale, and so are many railway, highway, and water conservation projects.

For example, the Lanchow-Sinkiang Railway, which traverses Kansu and Sinkiang Provinces; the Paochi-Chengtu Railway connecting Northwest and Southwest China; the Yingtan-Amoy Railway linking Kiangsi and Fukien Provinces; and the Chining-Erhlien Railway linking China, the Mongolian People's Republic, and the Soviet Union, which are being built in the present five-year period, not only have a long over-all mileage but also involve stupendous engineering feats in crossing deserts and towering mountain ranges.

The Sikang-Tibet and Chinghai-Tibet highways, construction of which was carried on in the present five-year period, were opened to traffic in 1954 and have a total length of more than 4 300 kilometres. They cut through mountain ranges rising several thousand metres above sea level, where there is scarcely a trace of human habitation. The engineering work involved was particularly difficult and massive in scale.

The project for the complete harnessing of the Huai River, which has been going ahead in the present five-year period, provides for four big reservoirs at Nanwan, Poshan, Futseling, and Meishan. These will be capable of storing more than 3 800 million cubic metres of water. At the same time, flood control and measures to deal with waterlogging will be carried out on the main tributaries of the Huai – the Hungho, Juho, Suiho, and Peifei rivers. The Kuanting Reservoir completed in 1954, with a storage capacity of about 2 300 million cubic metres of water, will play an important role in preventing floods on the lower reaches of the Yungting.

The permanent control of the Yellow River and multiple-purpose development of its resources will begin in the present five-year period. The Yellow River flows for more than 4 800 kilometres through seven provinces, with a drainage area of 745 000 square kilometres. In the past, it has caused more damage than any other river in the country. According to the master plan for its multiple-purpose utilization, dozens of dams will be built on its middle and lower reaches and on its main tributaries. Huge reservoirs capable of regulating its flow and big hydroelectric power stations will be erected at the Sanmen Gorge and four other sites. The work of drawing up the master plan for the permanent control of the Yellow River and multiple-purpose development of its resources will be completed in the period of the First Five-Year Plan, and construction will begin on the river regulation and hydroelectric power installations at the Sanmen Gorge.

Appendix 3

Second Five-Year Plan, 1956

Source

An extract from Xinhua News Agency release dated 28 September 1956.

The central task of our Second Five-Year Plan is still to give priority to the development of heavy industry. This is the chief index of our country's socialist industrialization, because heavy industry provides the basis for a strong economy and national defence, as well as the basis for the technical reconstruction of our national economy.

It is required that the output of the main products of heavy industry reach approximately the following levels in 1962:

Product	Target for 1962	Target for 1957	Actual Output in 1952	Peak Annual Output Before Liberation	
				Year	Output
Electric power (billion kwh.)	40–43	15.9	7.26	1941	5.96
Coal (million tons)	190–210	112.985	63.528	1942	61.875
Crude oil (million tons)	5–6	2.012	.436	1943	.32
Steel (million tons)	10.5–12	4.12	1.35	1943	.923
Aluminium ingots (thousand tons)	100–120	20.0	—	—	—
Chemical fertilizers (million tons)	3–3.2	.578	.194	1941	.227
Metallurgical equipment (thousand tons)	30–40	8.0	—	—	—
Power-generating equipment (million kw.)	1.4–1.5	.164	.0067	—	—
Metal-cutting machine tools (thousand units)	60–65	13.0	14.0	1941	5.0
Timber (million cubic metres	31–34	20.0	10.02	—	—
Cement (million tons)	12.5–14.5	6.0	2.86	1942	2.293

In the Second Five-Year Plan period, we must make vigorous efforts to expand the machine-building industry, particularly that making industrial equipment, and continue to expand the metallurgical industry to meet needs of national construction. At the same time, we should also energetically develop the electric power, coal-mining, and building materials industries and strengthen the backward branches of industry – the oil, chemical, and radio equipment industries. We should press ahead vigorously with the establishment of industries utilizing atomic energy for peaceful purposes.

No effort should be spared in this five-year period in strengthening the weak links in our industry and in opening up new fields, such as the manufacture of various kinds of heavy equipment, machine tools for special purposes, and precision machine tools and instruments; the production of high-grade alloy steels; the cold working of steel products; the mining and refining of rare metals; ... the setting up of an organic synthetic chemical industry; and so forth. At the same time, we should also pay attention to multiple-purpose utilization of resources, particularly the over-all use of the associated nonferrous metals.

Appendix 4

Fourth and Fifth Five-Year Plan

Source

Adapted in parts from *China's Petroleum Industry*, National Council for US-China Trade, Washington DC 1976, pp 16–19

Fourth FYP, 1971–75 (first paragraphs);
Fifth FYP, 1976–80 (second paragraphs)

Agriculture

Grain output rose 12.5 per cent in 4th FYP from 240 m tons in 1970 to about 270 m tons in 1975; annual growth was erratic – +2.5 per cent in 1971, −2.5 per cent in 1972, +4.2 per cent in 1973, +6 per cent in 1974, and +2 per cent in 1975. Average: 2.3 per cent pa. Contributing were 64 per cent increase in chemical fertilizer use, new irrigation wells, enlarged irrigated area by 6.5 m ha and improved drainage on another 6.5 m ha during 1971–75. New seed, development yet to show results; general research techniques still 25 years behind West.

Ambitious 400 m mt goal set for 1980 grain and soybean output requiring nearly 7 per cent yearly growth 1976–80. Planners will rely on mechanization, fertilizer, and improved management. But returns from new inputs may not be realized until complementary factors in place. Irrigation systems possible bottleneck. 300 m mt goal more realistic with agricultural takeoff in 1980s. (274.9 m tons output figure for 1974 announced by the Chinese refers probably to grain plus soybeans: 255–280 grain, 15–20 soybeans.)

Agricultural Mechanization

Agricultural mechanization drive, initiated 1973, led to large production increases. Est. 1975 output of 54 000 standard tractors and 42 420 walking tractors, up 92 per cent over 1971. Total horsepower of irrigation systems: 37 599 000 in 1975. 87 per cent over 1971 capacity. Production organization has machinery over 20 hp and engines over 4 cylinders manufactured at national level with smaller equipment at county and city plants. Communes adjust all machinery for local use. Mechanization power sources diesel for mobile units and electricity for stationary.

Mechanization expected to accelerate, with goal of total mechanization by 1980. Local efforts will be emphasized to develop additional semi-mechanized equipment, plus industrial efforts to mass produce rice transplantors and combine harvesting machinery. Priorities in order are water conservation and irrigation, food and fodder processing, threshing, land preparation, paddy transplanting, and harvesting of crop. Average Chinese farm machine works 2 500 hours a year, 8 times Japanese rate, 2.5 times US rate.

Steel

Crude steel production grew at promising 12.7 per cent pa rate until 1973 when stagnation of industry prevented further increase. 1975 crude steel output at about 25 m mt giving average 4th FYP yearly growth of 7.8 per cent. 1975 finished steel production was approximately 19.2 m mt, 43 per cent over 1970 figures. Past failure to expand mining and ore benefication operations caused unprecedented pig iron imports, over 1 m mt pa after 1973. Scrap supply and coking coal capacities also failed to keep pace. Steel finishing facilities have also been a bottleneck, but $500 million steel plant purchases from Japan and Germany will alleviate this.

Domestic finished steel requirements projected to reach 32.8 m mt by 1980, but current trends put 1980 production at 27 mt, with crude steel output at 35.9 m mt. Swing to BOFs. Probable program to relieve mining bottlenecks with new ore benefication systems and other primary operations. Balance could be regained by 1980 in time for 6th FYP.

Chemicals and Petrochemicals

1975 chemical fertilizer production of 28.7 m mt. double 1970 output. 4th FYP saw construction of 28 large and medium scale fertilizer plants. Intermediate plants bought abroad 1971–75 for ethylene-glycol, polypropylene, ammonia, titanium trichloride, butadiene, and benzene, at total cost: $1 billion. 1975 synthetic fibre production est. 85 per cent above 1970 output, insecticides up 77 per cent, pharmaceuticals up 90 per cent. Small scale plant output up, eg, small synthetic ammonia plants accounted for 54 per cent of output by 1973.

Foreign plants coming on stream through 5th FYP. At least 13 plants with fertilizer capacities of 330 000 mt pa each will double nitrogen supply to est. 28.7 m mt by 1980. Petrochemical plants to start production of urea, DMT, polyester chips, cyclohexane, aromatic extraction, and catalysts. Development of new cracking techniques for acetone, isopropyl alcohol, CIS-1, 4-polybutidiene, polyethylene, polypropylene, and acrylonitrile production should complement intermediate petrochemical facilities purchased during 4th FYP.

Nonferrous Metals and Minerals, Coal

Tungsten production steady at 15 000 tons pa through 4th FYP. Antimony down 25 per cent from 1965 output to 12–14 000 ton range due to low international price. Tin at about 20 000 ton annual production, also below potential output. Coal, however, up est. 5 per cent pa to 411 m tons in 1975, accounting for 69 per cent domestic energy supply. Imports of copper, aluminum, and nickel alone increased an average of 22.6 per cent pa during 1971–5, together costing 50 per cent more than all China's metal and mineral exports during same period.

Major efforts in coal mining are expected with yearly growth rate of over 8 per cent. 1980 production could surpass 603 m tons. Additional supplies may be needed to save oil for export and supply burgeoning steel industry. Tungsten, antimony and tin production will increase only with world price hike. Aluminum, nickel, copper to remain expensive import items.

Oil

Crude oil production rose 169 per cent from 28.5 m mt in 1970 to 76.7 m mt in 1975, an average annual growth rate of 21.9 per cent. Chinese recoverable reserves thought between 20 and 40 billion barrels, with est. ranging from 7 to 235 billion. China would rank 10th in world reserves with 20 billion, and 5th with 40. Known oil-related equipment sales to China of at least $340 m during 4th FYP.

Growth for 5th FYP est. at 17.7 per cent pa pushing output to 173.3 m tons in 1980, a 125 per cent jump, Investment required in petroleum infrastructure during 1976–80 could be equivalent of $5 billion at $50 per ton of increased output. Purchases abroad for harbour facilities, refining and transportation systems will be placed 1976–78.

Power Generation

Electric power generated in China has increased est. 65 per cent from 72 billion kW in 1970 to 125 kw in 1975, with average growth rate of 11.8 per cent pa. Current level of technology allows serial production of 400 ton per hour steam boilers, 125 MW stream turbines and 72.5 MW hydro turbines, and 110/120 kV transmission line. Installed generating capacity grew 39 per cent during 4th FYP to 36 700 MW in 1975. Development of 300 000 kW thermal generator reported. Sources of energy in 1975: coal 69 per cent, oil 20 per cent, gas 10 per cent, hydro 1 per cent.

76 per cent increase expected in electric power production pushing output to 221 billion kW by 1980, with demand thought to increase at 12 per cent pa reaching 186 billion kW in 1980. Capacity will increase 64 per cent to 58 600 MW by 1980. Development during 6th FYP should centre around 670 mt/hr boilers, 200 and 300 MW steam turbines, 150 MW hydro turbines and 330 kV transmission lines. Probable energy distribution in 1980: coal 57 per cent, oil 30 per cent, gas 12 per cent, hydro 1 per cent.

Electronics

Radio production est. 15 m units in 1975, nearly 300 per cent above 1970 output, surpassed only by US and Japan. New emphasis on TV production reaching approximately 191 000 units in 1975 compared to 1970s 15 000. Colour TV output began. Computer systems were developed during 4th FYP with a million operations per second capacity, production over 100 units pa. LSI unit components of 10 000 elements in production 1975. Telecommunication equipment was a top 4th FYP priority.

Mass production of television may follow current boom in radios; production could reach 1 m units pa by 1980. Potential need for 2 000 to 2 500 large computer systems will spur increased domestic production. Main priorities: colour television, high precision testing equipment, monitoring and control instruments for industrial purposes. Planned automation will increase demand for electronics. Small-scale plant output will expand.

Machine Tools

Marginal increases in production led to est. 1975 85 000 unit output, 21 per cent above the 1970 production of 70 000 units. Chinese technological advances of 4th FYP included simple numerically controlled machine tools, fully automated heavy duty thread grinders, and large column jig borers. Research done one reproduction of foreign manufacturing methods.

With slightly better manufacturing techniques, production in 1980 could pass 114 000 units pa. Chinese present know-how sufficient to handle general production requirements for agricultural mechanization programme and mining industry, but foreign expertise will be needed in specialty fields. Chinese capacities should allow export of smaller, less complex machine tools within 10 years.

Transportation

Expansion of economy and development of port facilities: 40 10 000 dwt berths built in last 3 years to increase cargo loading capacity by 50 per cent. Rail system grew 20 per cent during 4th FYP from 40 000 km in 1971 to 48 000 in 1975; first electrified line opened; rail freight moved in 1975 est. at 900 m mt, up 45 per cent from 620 m mt in 1970. Serious bottlenecks 1974. Surge in road vehicle output and use.

Expansion in all areas expected. Completion of at least one 100 000 dwt berth for larger tankers and further port development to match increased trade anticipated. New railroad network plus some double tracking and electrification of trunk line expected. Further expansion of road system predicted to reach 2 per cent of communes and 17 per cent of production teams still outside motor access.

Abreviations: FYP = Five Year Plan; ha = hectacre; hp = horsepower; kV = kilovolt; kW = kilowatt; dwt = dead weight tons; km = kilometre; lin = linear; m = million; mt = metric tons; MW = megawatts; pa = per annum, est = established.

Appendix 5

Sixth Five-Year Plan, 1981

Source

An extract from Premier Zhao Ziyang's Report at the Fifth National People's Congress, 30 November 1982, and printed in *Beijing Review* 20 December 1982.

Under the Sixth Five-Year Plan, total investment in capital construction throughout the country will be 230 billion yuan, roughly the same as for the Fifth Five-Year Plan. Altogether 890 large and medium-sized projects will be undertaken during these five years, far fewer than during the previous five-year period. This is based on the lessons of the past when our capital construction was too large in scale and overextended, so yielding poor returns from investment.

In the Sixth Five-Year Plan, energy and transport are priorities and will receive 38.5 per cent of the total investment in capital construction, a somewhat higher figure than before. Meanwhile, appropriate arrangements are made for development in agriculture, the textile and other light industries, the metallurgical industry and the chemical industry, as well as for education, science, culture, public health service, urban public utilities, commerce and foreign trade. Four hundred large and medium-sized projects are to be completed before the end of 1985, and the rest will be carried over for continued construction during the Seventh Five-Year Plan period.

Funds totalling 130 billion yuan are earmarked for updating equipment in the existing enterprises and for their technical transformation. In the 28 years between 1953 and 1980, such funds accounted for about 20 per cent of the total investment in fixed assets. The figure is raised to 36 per cent in the Sixth Five-Year Plan. This is a big step forward. It will help improve the technology of our industry and speed up the modernization of the national economy as a whole.

I would now like to speak at some length about the development plans for energy, transport and agriculture and about the priorities for the technical transformation of the existing enterprises.

The coal industry will receive 17.9 billion yuan in investment during the five years. The funds will be used chiefly for exploiting the coal reserves in Shanxi, northest China and eastern Inner Mongolia; at the same time, the coalfields in western Henan, Shandong, Anhui, Jiangsu and Guizhou Provinces will also be developed. Twenty-eight large coal mines will be added, each with an annual capacity of over one million tons. This, plus the opening of small and medium-sized mines, will increase the total capacity of China's coal industry by 220 million tons. The plan provides for a production capacity of 80 million tons to be available before the end of 1985, with work continuing during the next five-year plan period for the remaining 140 million tons.

To speed up the expansion of the coal industry, we should, on the one hand, concentrate on exploiting the big opencast mines and, first of all, on building as soon as possible five such mines at Huolinhe, Yiminhe, Pingshuo, Yuanbaoshan and Jungar, which are rich in coal deposits. On the other hand, we should step up the technical transformation of the existing mines, build small and medium-sized ones and tap potential. Thus we will be able to produce more coal with less investment at an earlier date, and we can increase our coal output appreciably in the next few years or within this decade and at the same time lay the ground for greater expansion of our coal industry in the following decade.

A total of 15.4 billion yuan will be invested in the petroleum industry in the Sixth Five-Year Plan period. Efforts will be concentrated on prospecting in the Songliao Basin of northeast China, the Bohai Bay, the Puyang Region of Henan Province and the Eren Basin of Inner Mongolia. General survey and prospecting will also be stepped up to a proper extent in the Junggar Basin of Xinjiang and the Qaidam Basin of Qinghai, and such work will be steadily continued for present oil and natural gas fields in east China. Our goal is to verify the reserves of a number of new oilfields and to actively prospect for and exploit offshore oil deposits. In these five years, production capacity for oil will increase by 35 million tons and for natural gas by 2.5 billion cubic meters. This should help make up for the depletion as a result of continued exploitation of oil wells now in operation and maintain our annual oil output level at 100 million tons during the 1981–85 period.

A total of 20.7 billion yuan will be invested in the power industry in the same period. It will be used chiefly for continued exploitation of hydraulic resources along the upper reaches of the Huanghe River and the upper and middle reaches of the Changjiang River and in the Hongshui River Basin, and for building a number of large hydroelectric stations. At the same time, a number of electric power stations will be constructed near the coal mines in coalrich Shanxi Province, eastern Inner Mongolia, Huainan and Huaibei regions, western Henan Province, areas north of the Weihe River and Guizhou Province, as will be a number of thermal power stations in Shanghai and Liaoning, Jiangsu, Zhejiang, Guangdong and Sichuan Provinces where enormous amounts of electricity are consumed.

These five years will witness the construction or continued construction of 15 hydroelectric stations each with an installed capacity of 400 000 kW or more, 45 thermal power stations each with an installed capacity of 200 000 kW or more, and one nuclear power station with an installed capacity of 300 000 kW. Added to the small stations to be set up, this will mean a total increase of 36.6 million kW of electric power for the whole country. Of this figure, 12.9 million kW will be available before the end of 1985, while work on the remaining 23.7 million kW will continue in the next five-year plan period. Since these arrangements still fall short of fully meeting the power requirements of China's economic growth, efforts will be made, wherever possible, to step up construction of electric power stations in the course of implementing the plan, while strictly economizing on the use of electricity.

Twenty-nine point eight billion yuan will be invested in transport and post and telecommunications services over the five years, mainly for railway and harbour construction. The Sixth Five-Year Plan provides for laying 2 000 kilometres of rails, double-tracking 1 700 kilometres of the present railways and electrifying 2 500 kilometres. By 1985, the capacity of transporting coal from Shanxi, western Inner Mongolia and Ningxia to other places will have increased from 72 million tons in 1980 to 120 million tons and that to northeast China from 14 million tons in 1980 to 29 million tons.

Construction of 132 deepwater berths is scheduled for 15 harbours including Dalian, Qinhuangdao, Tianjin, Qingdao, Shijiusuo, Lianyungang, Shanghai, Huangpu and Zhanjiang so that the handling capacity of the country's coastal harbours will total 317 million tons by 1985 as against 217 million tons in 1980. Work will continue on inland navigation projects along the Changjiang and other rivers. More roads will be built so as to improve the rural transport conditions. In the meantime, efforts will be made to increase the post and telecommunications facilities. Completion of these tasks will relieve the strain on transport and post and telecommunications services.

In agriculture, stress will be put on augmenting the flood control capacity of the Huanghe, Changjiang, Huaihe and Haihe Rivers and on completing the construction of the Panjiakou and Daheiting reservoirs in Hebei Province and the Tianjin project for diverting water from the Luanhe River so as to ease the shortage of water for industrial and agricultural use in the Beijing-Tianjin area. Continued efforts will be made to build commodity grain bases on the Sanjiang Plain in Heilongjiang Province, in the Poyang Lake area in Jiangxi Province, the Dongting Lake area in Hunan Province and the Pi-Shi-Hang Irrigation Area in Anhui Province. A network for breeding and popularizing improved seeds should be set up and improved step by step.

Shelterbelt networks in northwest, north and northeast China will continue to be built so as to check soil erosion in the areas along the middle reaches of the Huanghe River and the shifting of sand dunes in the northwest. We should make sustained efforts to conduct the nationwide afforestation campaign in order to cover our motherland with trees. Land reclamation on the pasturelands of the minority nationality areas is forbidden. We should expand the total area of artificially sown pastures from the 32 million *mu* [One *mu* equals 1/15 hectare] in 1980 to 100 million *mu* in 1985. A number of farms producing good poultry and animal strains, livestock farms and feed-processing plants will be set up. Between 1981 and 1985, the plan is to increase the area for fresh-water acquaculture by 16 million *mu* and that for seawater aquaculture by 800 000 *mu*. All this will help improve conditions for farming, animal husbandry and fishery.

We should do more geological prospecting, hunt for mineral deposits, assess natural resources and do better work in hydrogeology, engineering geology and environmental geology during the Sixth Five-Year Plan period.

In the technical transformation of the existing enterprises stress should be put on the following: saving energy and raw and semi-finished materials; improving the product mix; enhancing the properties and quality of products; and increasing the production capacity for certain urgently needed products which are in short supply. Measures will be taken in these five years to carry out in a planned way the technical transformation of a number of key enterprises such as the Anshan Iron and Steel Company, the Baotou Iron and Steel Company, the Changchun No. 1 Motor Works, factories in the power equipment manufacturing bases in Harbin and Shanghai, the large Datong, Kailuan and Fushun coal mines and the large caustic soda plants in Tianjin and in Hubei, Liaoning, Sichuan and Shandong Provinces. The purpose is to gradually raise the technological level of these key enterprises that have bearing on the national economy as a whole.

As the supplier of equipment for diverse fields, the machine-building industry must be ahead of others in carrying out technical transformation. In the 1981–85 period, mechanical and electrical products that are mass produced and widely used, especially motor vehicles, tractors, internal-combustion engines and industrial boilers that consume much energy should be improved technically and updated. Research and development of a group of key equipments such as precision and highly efficient machine tools, instruments and metres should be undertaken, and efforts should be made to update many kinds of instructions, components and basic parts.

We should turn out equipment of a fairly high technical level for farming, animal husbandry and fishery as well as for the textile and other light industries. We should also strive to improve the technology of manufacturing complete sets of large equipment used in power generation, steel rolling, mining, coal washing, transport, offshore oil prospecting and the petrochemical industry. Success in these endeavours will mean a gratifying improvement in the technological level of our national economy.

Constantly raising the educational, scientific, technological and cultural levels of the whole people is a major guarantee for building a modern material civilization and also a major aspect in building a socialist spiritual civilization. In the Sixth Five-Year Plan, allocations for education, science, culture, public health and physical culture account for 15.9 per cent of total state expenditure (the corresponding figure will reach 16.8 per cent in 1985), a fairly big increase over the 11 per cent of the Fifth Five-Year Plan. The funds allocated for these undertakings are admittedly still insufficient, but this is what we can afford at present in the context of our limited financial resources. Appropriations for these undertakings will gradually increase along with the future growth of our economy.

In education the Sixth Five-Year Plan calls for a rise in the number of newly admitted full-time students in institutions of higher education from 280 000 in 1980 to 400 000 in 1985, an increase of 42.2 per cent, while total enrolment is to reach 1.3 million, 13.6 per cent more than in 1980. The number of graduates from these institutions is to be 1.5 million for the five years covered by the plan. Meanwhile, there will be considerable expansion of higher education through radio, TV, correspondence and evening courses. In 1985, a total of 20 000 postgraduate students is to be admitted, 5.5 times the number in 1980, and 45 000 are to complete their postgraduate studies in the five years under review.

The state will appropriate special funds for the construction of fairly advanced teaching and laboratory facilities in a group of key colleges and universities, and for the building or expansion of a number of experimental centres so as to raise the standard of instruction in these institutions. Specialities in colleges and universities will be readjusted and teaching methods improved. Over the years, specialities have been too finely divided and consequently students can acquire only a limited scope of knowledge. This falls short of the requirements of practical work in various fields and for advanced studies and often creates difficulties for the graduates to get employment or to switch to other specialized fields of work. This situation must be changed. Systematic education of our undergraduates and postgraduates in the basic theories of Marxism must be stepped up and ideological and political work among them should be done regularly, purposefully and effectively. We must ensure that our students are trained to be educated workers with socialist consciousness and professional knowledge.

We must continue to restructure secondary education, set up secondary vocational schools of different types, particularly for farming, forestry, animal husbandry, sideline occupations and fishery, the medical and nursing services, finance and trade, public security and procuratorial and judicial affairs, and culture and education. Some vocational and technical subjects are to be added to the curricula of regular secondary schools. It is essential to create the necessary conditions for transforming a number of regular senior middle schools in the rural areas into secondary vocational schools serving agriculture. More attention should be paid to pre-school education and teachers' training.

We should strive to make primary school education universal or almost universal by 1985 in most counties and to make junior middle school education universal in the cities. Schools of all types and levels should try to enhance the professional competence of the teachers and their teaching level in an all-round way and gradually to improve the conditions for both teaching and learning. Educational workers throughout the country face the most arduous and pressing task of making primary and junior middle school education universal and eliminating illiteracy among young and middle-aged people, and the whole nation should go into action and give this work full support. We should make respect of educational work a standard of good social conduct so that everybody will realize that it is the very foundation of our modernization drive.

In science and technology, we will, under the Sixth Five-Year Plan, put into nationwide use the verified results of 40 key scientific and technological research projects in agriculture, the textile and other light industries, energy, the electronic and machine-building industries, raw and semi-finished materials, the chemical industry, the pharmaceutical industry, transport, and post and telecommunications. To meet the needs of economic and social development, we should tackle 100 major problems in 38 scientific and technological research projects vital to production and construction, and try to bring a considerable number to fruition and widely apply the results during this period. While focusing on research in development and application, we should step up basic research so as to provide China's scientific and technological development with sound guidance and a reliable foundation. In line with the characteristics of the specific disciplines, research in both basic and applied sciences should be geared as much as possible to the needs of economic development and expanding production. We should put the application and dissemination of the results of scientific research on a par with research itself, commend and reward successes in this field and overcome the tendency to underrate its importance. It is necessary to set up, in a planned way and step by step, public centres of information, forecasting and analysis, measurement, calculation and consulting services in applied mathematics in the provinces, municipalities and autonomous regions to serve scientific research and growth of the national economy.

While energetically expanding research in the natural sciences, we should attach importance to studies in the social sciences, in political economy, departmental economics, management science, philosophy, law, political science, education, sociology, ethics, psychology, history, ethnology, literature and art, linguistics, international relations, and so on. We should do our best to expound and solve the major theoretical and practical problems raised in the course of our socialist modernization so that we can use the results of creative studies to help build the socialist material and spiritual civilization and develop our socialist system. Gaps in our social studies should be filled in; weak links strengthened and working conditions improved.

Appendix 6

Perspectives to the Seventh Five-Year Plan, 1986

Source

An excerpt from Chairman Hu Yaobang's Report to the Twelfth Party Congress, 1 September 1982 and printed in *Beijing Review*, 13 September 1982.

The general objective of China's economic construction for the two decades between 1981 and the end of this century is, while steadily working for more and better economic results, to quadruple the gross annual value of industrial and agricultural production – from 710 billion yuan in 1980 to 2 800 billion yuan or so in 2000. This will place China in the front ranks of the countries of the world in terms of gross national income and the output of major industrial and agricultural products; it will represent an important advance in the modernization of her entire national economy; it will increase the income of her urban and rural population several times over; and the Chinese people will be comparatively well-off both materially and culturally. Although China's national income per capita will even then be relatively low, her economic strength and national defence capabilities will have grown considerably, compared with what they are today. Provided that we work hard and in a down-to-earth manner and bring the superiority of the socialist system into fuller play, we can definitely attain our grand strategic objective.

From an overall point of view, what is most important in our effort to realize this objective in economic growth is to properly solve the problems of agriculture, energy and transport and of education and science.

Agriculture is the foundation of the national economy, and provided it grows, we can handle the other problems more easily. At present, both labour productivity and the percentage of marketable products are rather low in our agriculture; our capacity for resisting natural calamities is still quite limited; and, in particular, the contradiction between the huge population and the insufficiency of arable land is becoming ever more acute. From now on, while firmly controlling the population growth, protecting all agricultural resources and maintaining the ecological balance, we must do better in agricultural capital construction, improve the conditions for agricultural production, practise scientific farming, wrest greater yields of grain and cash crops from limited acreage, and secure the all-round development of forestry, animal husbandry, sideline occupations and fishery in order to meet the needs of industrial expansion and of higher living standards for the people.

Energy shortage and the strain on transport are major checks on China's economic development at present. Growth in energy production has slowed down somewhat in the last few years, while waste remains extremely serious. Transport capacity lags far behind the increasing volume of freight, and postal and telecommunications facilities are outmoded. To ensure a fair rate of growth in the national economy, it is imperative to step up the exploitation of energy resources, economize drastically on energy consumption and at the same time strive hard to expand the transport and postal and telecommunications services.

The modernization of science and technology is a key link in our four modernizations. Today, many of our enterprises are backward in production techniques, operation and management; large numbers of workers and staff members lack the necessary scientific knowledge, general education and work skills; and there is an acute shortage of skilled workers, scientists and technicians. In the years to come, we must promote large-scale, technical transformation in a planned way, popularize technical measures that have yielded good economic results, and actively introduce new techniques, equipment, technologies and materials. We must step up research in the applied sciences, lay more stress on research in the basic sciences and organize people from all relevant fields to tackle key problems in scientific research. We must improve our study and application of economics and scientific business management and continuously raise the level of economic planning and administration and of the operation and management of enterprises and institutions. And we must work vigorously to universalize primary education, strengthen secondary vocational education and higher education and develop educational undertakings of all types and at all levels in both urban and rural areas, including training classes for cadres, workers, staff members and peasants and literacy classes in order to train all kinds of specialists and raise the scientific and educational level of the whole nation.

In short, in the next 20 years we must keep a firm hold on agriculture, energy, transport, education and science as the basic links, the strategic priorities in China's economic growth. Effective solution of these problems on the basis of an overall balance in the national economy will lead to a fairly swift rise in the production of consumer goods, stimulate the development of industry as a whole and of production and construction in other fields and ensure a betterment of living standards.

Population has always been an extremely important issue in China's economic and social development. Family planning is a basic policy of our state. We must do our utmost to keep our population within 1.2 billion by the end of this century. The total number of births is now at its peak. Excessive population growth will not only adversely affect the increase of percapita income but also cause serious difficulties in food supply, housing, education and employment, and it may even disrupt social stability. Consequently, we must never slacken our effort in family planning, especially in the rural areas. We must conduct intensive and meticulous ideological education among the peasants. Provided that we do our work well, we can succeed in bringing our population under control.

In order to realize our objective for the next two decades, we must take the following two steps in our strategic planning: in the first decade, aim mainly at laying a solid foundation, accumulating strength and creating the necessary conditions; and in the second, usher in a new period of vigorous economic development. This is a major policy decision taken by the Central Committee after a comprehensive analysis of the present conditions of China's economy and the trend of its growth.

Our national economy has grown steadily even in the past few years of readjustment, and the achievement is quite impressive. In many fields, however, the economic results have been far from satisfactory, and there has been appalling waste in production, construction and circulation. We have yet to equal our best past records in the materials expended in per unit products, in the profit rate of industrial enterprises, in the construction time for large and medium-sized projects and in the turnover rate of circulating funds in industrial and commercial enterprises. Apart from some objective factors not subject to comparison, the main causes for this are the 'Left' mistakes of the past, which resulted in blind proliferation of enterprises, an irrational economic structure, defective systems of economic administration and distribution, chaotic operation and management and backward production techniques. Things started to pick up a little in 1982, with the stress laid on better economic results. Nevertheless, it is impossible in a brief space to solve all such problems which have piled up over a long period. We have to bear this basic fact in mind when drawing up the strategic plan for China's economic development.

In the period of the Sixth Five-Year Plan (1981–85), we must continue unswervingly to carry out the principles of readjustment, restructuring, consolidation and improvement, practise strict economy, combat waste and focus all economic work on the attainment of better economic results. We must devote our main efforts to readjusting the economic structure in various fields, streamlining, reorganizing and merging the existing enterprises and carrying out technical transformation in selected enterprises. At the same time, we must consolidate and perfect the initial reform in the system of economic administration and work out at an early date the overall plan for reform and the measures for its implementation. During the Seventh Five-Year Plan (1986–90), we shall carry out the technical transformation of enterprises on an extensive scale and gradually reform the system of economic administration, in addition to completing the rationalization of the organizational structure of enterprises and the economic structure in various fields. We must also undertake a series of necessary capital construction projects in the energy, transportt and some other fields, and the concentrated solution of a number of major scientific and technological problems in the 1980s. Therefore, it will not be possible for the national economy to develop very fast in this decade. But if we complete the above tasks, we can solve the problems left over from the past and build a relatively solid basis for economic growth in the decade to follow. The 1990s will witness an all-round upsurge in China's economy which will definitely grow at a much faster rate than in the 1980s. If we publicize and explain this strategic plan adequately to the people, they will see the bright future more clearly and be inspired to work with greater drive to usher in the new period of vigorous economic growth.

Appendix 7

1978–85 Science Plan

Source

An excerpt from Vice-Premier Fang Yi's speech at the All-China Science Conference, 18 March 1978 and printed in *Beijing Review*, 7 April 1978.

Since last June, departments under the State Council and various localities and units have, through repeated discussions and revisions, mapped out a draft outline National Plan for the Development of Science and Technology, 1978–85.

The draft outline plan sets forth the following goals to be attained in the next eight years:

(1) Approach or reach the advanced world levels of the 1970s in a number of important branches of science and technology.

(2) Increase the number of professional scientific researchers to 800 000.

(3) Build a number of up-to-date centres for scientific experiment.

(4) Complete a nationwide system of scientific and technological research.

The eight-year outline plan (draft) makes all-around dispositions for the tasks of research in 27 spheres, including natural resources, agriculture, industry, national defence, transport and communication, oceanography, environmental protection, medicine, finance and trade, culture, and education, in addition to the two major departments of basic and technical sciences. Of these, 108 items have been chosen as key projects in the nationwide endeavour for scientific and technological research. When this plan is fulfilled, our country will approach or reach the advanced world levels of the 1970s in a number of important branches of science and technology, thus narrowing the gap to about ten years and laying a solid foundation for catching up with or surpassing advanced world levels in all branches in the following fifteen years.

The eight-year outline plan (draft) gives prominence to the eight comprehensive scientific and technical spheres, important new techniques, and pace-setting disciplines that have a bearing on the overall situation. It calls for concentrating all forces and achieving remarkable successes so as to promote the high-speed development of science and technology as a whole and of the entire national economy.

Agriculture. In accordance with the principle of 'taking grain as the key link and ensuring an all-around development,' we will in the next three to five years actively carry out comprehensive surveys of our resources in agriculture, forestry, animal husbandry, sideline production, and fisheries, study the rational exploitation and utilization of the resources and the protection of the ecological system, and study the rational arrangement of these undertakings.

We should implement in its entirety the Eight-Point Charter for Agriculture (soil, fertilizer, water conservancy, seeds, close planting, plant protection, field management, and improved farm tools) and raise our level of scientific farming. We should study and evolve a farming system and cultivating techniques that will carry forward our tradition of intensive farming and at the same time suit mechanization; and study and manufacture farm machines and tools of high quality and efficiency. We will study science and technology for improving soil, controlling water, and making our farmland give stable and high yields. In order to improve as quickly as possible the low-yielding farmland that accounts for about one-third or more of the country's total, we must make major progress in improving alkaline, lateritic, clay, and other kinds of poor soil, in preventing soil erosion, and in combating sandstorms and drought. We will study projects for diverting water from the south to the north and the relevant scientific and technical problems; study and develop new compound fertilizers and biological nitrogen fixation, methods of applying fertilizer scientifically, and techniques for drainage and irrigation; breed new seed strains, work out new techniques in seed breeding, and improve the fine crop varieties in an all-around way so that they will give still higher yields, produce seeds of better quality, and can better resist natural adversities. We should quickly develop new insecticides that are highly effective and are harmless to the environment, and devise comprehensive techniques for preventing and treating different kinds of plant diseases and pests.

We need to step up scientific and technological research in forestry, animal husbandry, sideline production, and fisheries. We should provide new tree varieties and techniques that will make the woods grow fast and yield more and better timber; develop multipurpose utilization of forest resources, and study techniques and measures for preventing and extinguishing forest fires; step up research on building pasturelands, improving breeds of animals and poultry, mechanizing the processes of animal husbandry, increasing the output of aquatic products, breeding aquatic products, and marine fishing and processing.

We will set up up-to-date centres for scientific experiments in agriculture, forestry, animal husbandry, and fisheries.

We must lay great emphasis on research in the basic theories of agricultural science, and step up our studies of the applications of agricultural biology, agricultural engineering, and other new techniques to agriculture.

Energy. We must make big efforts to accelerate the development of energy science and technology so as to carry out full and rational exploitation and utilization of our energy resources.

We have our own inventions in the science and technology of the oil industry, and in some fields we have caught up with or surpassed advanced levels in other countries. We must continue our efforts to catch up with and surpass advanced world levels in an all-around way. We should study the laws and characteristics of the genesis and distribution of oil and gas in the principal sedimentary regions, develop theories of petroleum geology, and extend oil and gas exploration to wider areas; study new processes, techniques, and equipment for exploration and exploitation and raise the standards of well drilling and the rate of oil and gas recovery; and actively develop crude oil processing techniques, use the resources rationally and contribute to the building of some ten more oilfields, each as big as Taching.

China has extremely rich resources of coal, which will remain our chief source of energy for a fairly long time to come. In the next eight years, we should mechanize the key coal mines, achieve complex mechanization in some of them, and proceed to automation. The small- and medium-sized coal mines should also raise their level of mechanization. Scientific and technical work in the coal industry should centre around this task, with active research in basic theory, mining technology, technical equipment, and safety measures. At the same time research should be carried out in the gasification, liquefaction, and multipurpose utilization of coal, and new ways explored for the exploitation, transportation, and utilization of different kinds of coal.

To improve the power industry is another pressing task. We should take as our chief research subjects the key technical problems in building large hydroelectric power stations and thermal power stations at pit mouths, large power grids, and super-high-voltage power transmission lines. We must concentrate our efforts on comprehensive research in the techniques involved in building huge dams and giant power-generating units and in geology, hydrology, meterology, reservoir-induced earthquakes, and engineering protection, which are closely linked with large-scale key hydroelectric power projects.

New sources of energy should be explored. We should accelerate our research in atomic power generation and speed up the building of atomic power plants. We should also step up research in solar energy, geothermal energy, wind power, tide energy, and controlled thermonuclear fusion, pay close attention to low-calorie fuels, such as bone coal, gangue, and oil shale, and marsh gas resources in the rural areas, and making full use of them where possible.

Attention should be paid to the rational utilization and saving of energy, such as making full use of surplus heat, studying and manufacturing fine and efficient equipment for this purpose, and lowering energy consumption by every means and particularly coke consumption in iron smelting, coal consumption in power generation, energy consumption in the chemical and metallurgical industries.

Materials.   Steel must be taken as the key link in industry. It is imperative to make a breakthrough in the technology of intensified mining and solve the scientific and technological problems of beneficiating hematite. We should speed up research work on the paragenetic deposits at Panchihhua, Paotow, and Chinchuan, where many closely associated metals have been formed; solve the major technical problems in multipurpose utilization; intensify research on the exploitation of copper and aluminium resources; make China one of the biggest producers of titanium and vanadium in the world; and approach or reach advanced world levels in the techniques of refining copper, aluminium, nickel, cobalt, and rare-earth metals. We should master new modern metallurgical technology quickly, increase varieties, and improve quality; study and grasp the laws governing the formation of high-grade iron ore deposits and the methods of locating them; establish a system of ferrous and non-ferrous materials and extend it in the light of the characteristics of our resources.

We should make full use of our rich natural resources and industrial dregs and increase at high speed the production of cement and new types of building materials which are light and of high strength and serve a variety of purposes; step up research in the technology of mining and dressing nonmetal ores and in the processing techniques; lay stress on research in the techniques of organic synthesis with petroleum, natural gas, and coal as the chief raw materials; step up our studies of catalysts and develop the technology of direct synthesis; renovate the techniques of making plastics, synthetic rubber, and synthetic fibre; and raise the level of equipment and automation in the petrochemical industry. We must solve the key scientific and technical problems in producing special materials necessary for our national defence industry and new technology, and evolve new materials characteristic of China's resources.

We should devote great efforts to basic research on the science of materials, develop new experimental techniques and testing methods, and gradually be able to design new materials with specified properties.

Electronic computers.   China must make a big new advance in computer science and technology. We should lose no time in solving the scientific and technical problems in the industrial production of large-scale integrated circuits, and make a breakthrough in the technology of ultra-large-scale integrated circuits. We should study and turn out giant computers, put a whole range of computers into serial production, step up study on peripheral equipment and software of computers and on applied mathematics, and energetically extend the application of computers. We aim to acquire by 1985 a comparatively advanced force in research in computer science and build a fair-sized modern computer industry. Microcomputers will be popularized, and giant ultra-high-speed computers put into operation. We will also establish a number of computer networks and data bases. A number of key enterprises will use computers to control the major processes of production and management.

Lasers.   We will study and develop laser physics, laser spectroscopy, and nonlinear optics in the next three years. We should solve a series of scientific and technical problems in optical communications, raise the level of the routine laser quickly, and intensify our studies of detectors. We expect to make discoveries and creations in the next eight years in exploring new types of laser devices, developing new laser wavelengths, and studying new mechanisms of laser generation, making contributions in the application of the laser to studying the structure of matter. We plan to build experimental lines of optical communications and achieve big progress in studying such important laser applications as separation of isotopes and laser-induced nuclear fusion. Laser technology should be popularized in all departments of the national economy and national defence.

Space.   We should attach importance to the study of space science, remote sensing techniques, and the application of satellites; build modern centres for space research and systems for the application of satellites; step up the development of the vehicle series, and study, manufacture, and launch a variety of scientific and applied satellites; actively carry out research in the launching of skylabs and space probes; and conduct extensive research in the basic theory of space science and the application of space technology.

High energy physics.   We expect to build a modern high-energy physics experimental base in ten years, completing a proton accelerator with a capacity of 30 000 million to 50 000 million electron volts in the first five years and a giant one with a still larger capacity in the second five years.

We should from now on set about the task in real earnest and make full preparations for experiments in high energy physics, with particular stress on studying and manufacturing detectors and training laboratory workers. We should step up research in the theory of high energy physics and cosmic rays; consciously promote the interpenetration of high energy physics and the neighbouring disciplines; actively carry out research in the application of accelerator technology to industry, agriculture, medicine, and other spheres; and pay attention to the exploration of subjects which promise important prospects of application.

Genetic engineering.   We must in the next three years step up the tempo of building and improving the related laboratories and conduct basic studies in genetic engineering. In the next eight years, we should combine them with the studies in molecular biology, molecular genetics, and cell biology and achieve fairly big progress. We should study the use of the new technology of genetic engineering in the pharmaceutical industry and explore new feasible ways to treat and prevent certain difficult and baffling diseases and evolve new high-yield crop varieties capable of fixing nitrogen.

Appendix 8

Directory of Major Research Establishments

The main research centres are listed below in alphabetical order under subject headings

Agriculture and Environmental Protection

Agrotechnology

Chinese Academy of Sciences
Changsha Institute of Agricultural Modernization
Dalian Institute of Chemical Physics
Heilongjiang Institute of Agricultural Modernization
Lanzhou Institute of Chemical Physics
Nanjing Institute of Soil Science
Northwest Institute of Soil and Water Conservation, Wugong
Qinghai Institute of Saline Lakes, Xining
Shijiazhuang Institute of Agricultural Modernization

Ministry of Chemical Industry
Shanghai Chemical Industry Research Institute
(see also p 78)

Ministry of Water Conservancy and Power
Water Conservancy and Hydroelectric Power Research Institute, Beijing

Crop Breeding

Chinese Academy of Sciences
Chengdu Institute of Biology
Institute of Botany, Beijing
Institute of Genetics, Beijing
Kunming Institute of Botany
Northwest Plateau Institute of Biology, Xining
Shanghai Institute of Plant Physiology
South China Institute of Botany, Guangzhou

Ministry of Agriculture Animal Husbandry and Fishery
Academy of Agricultural Sciences
Provincial Academies
(for details see pp 66–7)

Disease and Pest Control

Chinese Academy of Sciences
Institute of Entomology, Shanghai
Institute of Genetics, Beijing
Institute of Microbiology, Beijing
Institute of Zoology, Beijing
Kunming Institute of Zoology
Wuhan Institute of Virology

Ministry of Agriculture, Animal Husbandry and Fishery
(see under Crop Breeding)

Environmental Science

Chinese Academy of Sciences
Dalian Institute of Chemical Physics
Institute of Environmental Chemistry, Beijing
Institute of Geography, Beijing
Institute of Hydrobiology, Wuhan
Institute of Microbiology, Beijing
Nanjing Institute of Soil Science
Shanghai Institute of Organic Chemistry

State Oceanographical Bureau
Shangdong College of Oceanography, Quindao

Forestry and Desert Control

Chinese Academy of Sciences
Changchun Institute of Applied Chemistry
Lanzhou Institute of Desert
Shenyang Institute of Forestry and Soil Science
South China Institute of Botany, Guangzhou
Xinjiang Institute of Biology, Pedology and Psammology, Urumqi
Yunnan Institute of Tropical Botany, Mengla

Ministry of Forestry
Academy of Forest Science:
Institute of Forest Economics, Beijing
Institute of Forest Machinery, Yichundai Mountain (Heilongjiang)
Institute of Forest Products Chemical Industry, Nanjing
Institute of Forest Science, Beijing
Institute of Scientific and Technical Information for Forestry, Beijing
Institute of Subtropical Forestry, Fuyang county (Zhejiang)
Institute of Tropical Forestry, Jianfeng Mountain, Hainan Island (Guangdong)
Shellac Research Institute, Jingdong county (Yunnan)
Wood Industry Research Institute, Beijing
Provincial research establishments.

Biomedicine

Basic Studies

Chinese Academy of Sciences
Institute of Biochemistry, Shanghai
Institute of Biophysics, Beijing: Research departments in radiation biology, molecular biology, sensory receptors, submicroscopic cell structure, and technical projects and instrumentation
Institute of Cancer Research, Beijing
Institute of Cell Biology, Shanghai
Institute of Genetics, Beijing
Institute of Physiology, Shanghai
Institute of Zoology, Beijing

Cancer Research

Chinese Academy of Medical Sciences
Institute for Cancer Research, Ri Tan Hospital, Beijing
Institute of Cancer Research, Beijing
Institute of Haematology, Beijing Medical College
Shanghai First Medical College
Tianjin Institute of Tumours

Ministry of Public Health
Capital Hospital, Beijing

Chinese Traditional Medicine

Chinese Academy of Medical Sciences
Shanghai First Medical College

Chinese Academy of Sciences
Shanghai Institute of Physiology

Chinese Academy of Traditional Medicine
Beijing Traditional Medical College
Xiyuan Hospital, Beijing

Ministry of Public Health
Beijing Municipal Tuberculosis Research Hospital
Beijing Obstetrics and Gynaecology Hospital
Shanghai Sixth Hospital
Tianjin Municipal Hospital

Pharmacology

Chinese Academy of Medical Sciences
Institute of Materia Medica, Beijing

Chinese Academy of Sciences
Shanghai Institute of Biochemistry
Shanghai Institute of Materia Medica

Chinese Academy of Traditional Medicine
Institute of Traditional Chinese Pharmacology, Beijing

Surgery

Chinese Academy of Medical Sciences
Fuwai Hospital and Cardiovascular Institute, Beijing
Hsuanu Hospital and Neurosurgery Institute, Beijing
Institute of Haematology, Beijing Medical College
Institute of Organ Transplant, Wuhan Medical College
Ruijin Hospital, Shanghai Second Medical College
Zhongshan Hospital, Shanghai First Medical College
Zhongshan Medical College, Guangzhou

Ministry of National Defence
General Hospital of the People's Liberation Army, Beijing

Ministry of Public Health
Beijing Friendship Hospital
Beijing Jishutan Hospital
Shanghai Ninth People's Hospital
Shanghai Sixth People's Hospital

Earth Sciences

Climatology and Meteorology

Central Meteorological Bureau
Chengdu Institute of Meteorology
Institute of Climate, Beijing
Nanjing Institute of Meteorology
Linked to weather centres at provincial, municipal and local levels

Chinese Academy of Sciences
Commission for Integrated Survey of Natural Resources – Climate research group
Institute of Atmospheric Physics, Beijing
Lanzhou Institute of Plateau Atmospheric Physics

University of Beijing, Geophysics Department –
Meteorology Section

University of Nanjing, Meteorology Department

University of Science and Technology of China, Hefei
Earth and Space Sciences Department – Atmospheric Physics Section

Geochemistry

Chinese Academy of Sciences
Guangzhou Institute of Chemistry
Guiyang Institute of Geochemistry
Institute of Geology, Beijing
Institute of Oceanography, Qingdao
Qinghai Institute of Saline Lakes, Xining

State Oceanographical Bureau
Institutes of Oceanography at Hangzhou and Xiamen
Shandong College of Oceanography, Qingdao

University of Science and Technology of China, Hefei
Earth and Space Sciences Geochemistry Section

Geology and Geophysics

Chinese Academy of Sciences
Changsha Institute of Geotectonics
Institute of Acoustics, Seismoacoustics group
Institute of Geology, Beijing
Institute of Geophysics, Beijing

Ministry of Geology and Minerals
Colleges of Geology at Changchun, Zhangjiakou, Xian, Chengdu, Wuhan and Heifei
Institutes of Geology, Geomechanics, and Plateau Geology forming parts of the
Academy of Geological Sciences – (see p 122)
Linked to provincial bureaux of geology.

Nankai University, Tianjin, Geology Department

State Seismological Bureau
Factory No 581, Beijing
Institute of Engineering Mechanics, Harbin: earthquake engineering
Institutes of Geology and Geophysics, both in Beijing
Seismological Research Institute, Wuhan

University of Beijing, Geology Department;
Geophysics Department – Solid Earth Section (mainly geomagnetism and seismology)

University of Nanjing, Geology Department

Zhongshan University, Guangzhou, Geology Department
In addition to geology departments in ten other smaller universities.

Hydrogeology and Geomorphology

Chinese Academy of Sciences
Commission for Integrated Survey of Natural Resources –
Water Resources Division
Institute of Geography, Beijing
Institute of Glaciology and Cryopedology, Lanzhou
Northwest Institute of Soil and Water Conservation, Wugong
Provincial Institutes of Geography at Xinjiang, Urumqi, Changchun, Nanjing and
Chengdu

Ministry of Geology and Minerals
Institute of Hydrogeology and Engineering Geology, Zhengding
Institute of Karst Geology, Guilin

Ministry of Water Resources and Power
Institute of Hydrotechnical Research at Wukong and Zhengzhou
Water Conservancy and Hydroelectric Power Research Institute, Beijing

University of Beijing, Geography Department

Zhongshan University, Guangzhou, Geography Department
Over thirty other universities and colleges have geography departments, but the above
two are among those that are more active in basic research.

Oceanography

Chinese Academy of Sciences
Institute of Hydrobiology, Wuhan
Institute of Oceanography, Qingdao
South Sea Institue of Oceanography, Guangzhou

Ministry of Geology and Minerals
Geophysical Prospecting Bureau – Marine Division, Beijing
Petroleum and Marine Geological Bureau, Beijing

Shandong College of Oceanography, Qingdao

State Aquatic Products Bureau, (part of Ministry of Agriculture since 1982)
Fishery Institutes at Qingdao, Shangjai and Guangzhou
Research Laboratories run by provincial aquatic products bureaux.

State Oceanographical Bureau
    Institute of Marine Instrumentation, Tianjin
    Institutes of Oceanography at Hangzhou, Xuamen and Guangzhou
University of Xiamen, Oceanography Department

Energy Science and Technology

Chinese Academy of Sciences
    Chanchun Institute of Applied Chemistry: petroleum synthesis, solar cells
    Chengdu Institute of Biology, Bio-Energy Resources Division
    Chengdu Institute of Organic Chemistry: natural gas utilization
    Dalian Institute of Chemical Physics: liquid fuel synthesis, fuel cells
    Guangzhou Institute of Energy Conversion
    Hefei Institute of Plasma Physics
    Institute of Electrical Engineering, Beijing: magnetohydrodynamic generation, solar
    energy
    Institute of Engineering Thermophysics, Beijing: gas turbines
    Institute of Geology, Beijing: geothermal energy survey and exploitation
    Institute of Physics, Beijing: fusion (Plasma Physics Division) and solid-electrolytes
    for advanced batteries (Crystallography Division; see also Institute of Ceramic
    Chemistry and Technology below)
    Lanzhou Institute of Chemical Physics: petroleum fractioning
    Shanghai Institute of Ceramic Chemistry and Technology: solid-state batteries for
    intermittent energy storage or electricity generation load levelling and for cars
    Shanghai Institute of Nuclear Research
    Shanghai Institute of Optics and Fine Mechanics: laser initiation of fusion
    Shanxi Institute of Coal Chemistry, Taiyuan
    Southwest Institute of Physics, Dongshan: fission and fusion
    University of Science and Technology of China, Hefei, Department of
    Thermophysics

Ministry of Coal Industry
    Central Coal Research Institute, Beijing
    Changsha College of Coal Research
    Fushun College of Coal Research
    Institute of Coal Dressing Design, Beijing
    Institutes of Coal Chemistry at Beijing, Taiyuan and Yantai
    Tangshan College of Coal Research
    Other research laboratories run by provincial mining bureaux and major collieries.

Ministry of Machine Building Industry
    Institute of Transformer Research, Beijing
    Research departments attached to Xian, Paoding and Shengyang Transformer Plants

Ministry of Nuclear Energy
    First, Second, ... and Fifth Research Institutes
    Institute of Atomic Energy, Beijing
    Uranium Geology Research Institute, Beijing

Ministry of Petroleum Industry, Academy of Petroleum Research
    Colleges of Petroleum Engineering at Urumqi, Xian, Shenyang, Beijing and
    Nanchung (in Sichuan)
    Fushun Institute for Oil Shale Refining and Hydrogenation Research
    Harbin Institute of Petrochemistry
    Institute for Petrochemical Research, Beijing
    Institute of Crude Oil Refining, Beijing
    Institute of Petroleum, Beijing
    Institutes of Petroleum Research at Lanzhou, Dalian and Fushun
    Sichuan Institute of Natural Gas Research, Pingquan

Ministry of Water Conservancy and Power
 Hunan Institute of Water Conservancy and Hydroelectric Power Exploration and
 Design, Changsha
 Institute of Electrical Power Construction, Beijing
 Institute of Exploration Equipment, Shanghai: instruments used in geological survey
 for hydropower development
 Institute of Yangtze River Water Conservancy and Hydroelectric Power, Wuhan
 Institutes of Electrical Power Design at Xian, Wuhan and Shengyang
 Institutes of Hydroelectrical Engineering Design in Beijing and Shanghai
 Kuming Institute of Water Conservancy and Hydroelectric Power Exploration and
 Design
 Ministry of Water Conservancy and Power Research Institute, Nanjing
 Nanjing Institute of Hydraulic Engineering
 Shanghai Electrical Equipment Research Institute: especially transmission lines
 Tianjin Institute of Design
 Water Conservancy and Hydroelectric Power Research Institute, Beijing

State Energy Commission
 Institute of Energy Resources, Chengdu

Mineral Industries

Iron and Steel

Chinese Academy of Sciences
 Fujian Institute of Research on the Structure of Matter – Sanming branch
 Institute of Chemical Metallury, Beijing
 Shanghai Institute of Metallurgy
 Shenyang Institute of Metals Research

Ministry of Metallurgical Industry
 Changsha Institute of Mining and Metallurgy
 General Iron and Steel Research Institute, Beijing
 Northeast College of Technology, Shengyang
 Southwestern Iron and Steel Research Institute, Chengdu
 Taiyuan Iron and Steel Research Institute

Mineral geology and geophysical prospecting

Chinese Academy of Sciences
 Anhui Institute of Optics and Fine Mechanics, Hefei – Remote Sensing Division
 Changsha Institue of Geotectonics
 Commission for Integrated Survey of Natural Resources, Beijing – Energy Resources
 Department
 Guangzhou Institute of New Geological Techniques
 Guiyang Institute of Geochemistry
 Institute of Remote Sensing Application, Beijing
 Lanzhou Institute of Geology
 Wuhan Institute of Geodesy and Geophysics

Ministry of Coal Industry
 Central Coal Research Institute, Beijing
 Institute of Geology and Exploration, Beijing
 Xian Coal Field Geological Research Institute
 Xian Institute of Coal Geology for Coal Ash Analysis

Ministry of Geology and Minerals
  Academy of Geological sciences – 18 research institutes (see p 122)
  Aerogeophysical Prospecting and Xinjiang Aerial Survey Teams
  First and Second Marine Geological Survey Teams
  Geological Instrument Factories: Beijing, Shanghai, Xian, Chengdu and Chongqing
  Institute of Exploration Techniques, Beijing
  Institute of Geophysical Prospecting, Xian
  Institute of Geophysics and Geochemistry, Langfong, Hebei Province
  Research Groups on Ore Deposits Associated with Quaternary and pre-Cambrian
  Sedimentary Rocks, Tianjin; Ore Deposits Associated with Volcanic Rocks, Nanjing;
  and Sedimentary Ore Deposits, Chengdu
  Linked to provincial and municipal bureaux of geology, each with a number of field
  prospecting brigades, and to research institutes under the jurisdiction of provincial
  governments, eg, Shaanxi College of Metallurgical Prospecting and Design, Xian and
  Guilin Institute of Metallurgy and Geology.

Ministry of Metallurgical Industry
  Institute of Geology, Beijing

Ministry of Petroleum Industry
  Institute of Scientific Research for Petroleum Exploration, Beijing
  Sichuan Institute of Natural Gas Research, Pingquan

Ministry of Urban and Rural Construction and Environmental Protection
  Institute of Geology, Beijing

Mining science and technology

Chinese Academy of Sciences
  Wuhan Institute of Rock and Soil Mechanics – Statics of Rock Masses and of Soils
  divisions

Ministry of Coal Industry
  Central Coal Mining Research Institute, Beijing
  College of Coal Mine Design, Beijing
  Fuxin College of Coal Mining
  Mine Safety Instrument Development Institute, Fushun
  Shanghai Coal Mining Machinery Research Institute
  Shantung College of Mining, Jinan
  Sichuan College of Mining, Fuling (town where River Wu joins the Yangtze)

Ministry of Machine Building Industry
  Shenyang Institute of Heavy and Mining Machinery

Ministry of Metallurgical Industry
  Central–South Mining and Metallurgical College, Changsha
  General Mining and Metallurgical Research Institute, Beijing
  Shenyang Institute of Ore Dressing Machinery
  Tangsha Institute of Mining Research
  Xian Institute of Metallurgical Construction

Nonferrous metals

Chinese Academy of Sciences
  Changchun Institute of Applied Chemistry
  Institute of Chemical Metallurgy, Beijing
  Institute of Physics, Beijing – Laboratory of Physical and Chemical Analyses
  Shenyang Institute of Metal Research

Ministry of Metallurgical Industry
    Baoji Institute for Nonferrous Metals
    General Research Institute for Nonferrous Metals, Beijing
    Shanghai Institute of Materials Research
    Shanghai Institute for Nonferrous Metals

Exhaustive but confusing lists have been avoided in these sections relating to Mineral Industries. The research centres under various ministries and provincial authorities seem to present a picture of considerable fragmentation, duplication and mutual isolation in activities. R&D institutes run by individual mines and plants are too numerous to be included.

Transportation and Information Transformation

Aerospace

Chinese Academy of Sciences
    Guangzhou Institute of Electronic Technology
    Institute of Engineering Thermophysics, Beijing
    Institute of Mechanics, Beijing
    Lanzhou Institute of Chemical Physics
    Nanjing Astronomical Instrument Factory
    Space Science and Technology Centre, Beijing
    University of Science and Technology of China, Hefei, Modern Mechanics
    Department: high-temperature thermophysics

Ministry of Aviation Industry
    Aeronautical Institutes at Shenyang, Xian and Chengdu
    Colleges of Aeronautics at Harbin, Shenyang, Beijing, Zhengzhou, Nanjing, and Nanchang
    Turbojet Engine Research Centre at Jiangyou in Sichuan
    '303' and '625' Research Institutes in Beijing, etc.

Ministry of Space Industry
    Institute of Space Technology, Beijing

Quinghua University, Beijing, Engineering Mechanics Department: ionized gas dynamics

Xian Polytechnic University, Aircraft Engineering Department

Computers

Chinese Academy of Sciences
    Chengdu Institute of Computer Application
    Chengdu Scientific Instrument Factory
    Computer Centre, Beijing
    Harbin Institute of Precision Instruments
    Institute of Automation, Beijing
    Institute of Computer Technology, Beijing
    Shenyang Institute of Computer Technology
    Xinjiang Institute of Physics, Urumqi, Computer Application Division
    University of Science and Technology of China, Hefei, Radio and Electronics
    Department – Computer Section

Fudan University, Shanghai, Computer Science Department

Ministry of Electronics Industry
'1448' Research Institute, Changsha
Research Institute of Shanghai Instruments and Meters Plant

Qinghua University, Beijing, Computer Technology and Science Department

Shanghai Polytechnic University, Electronic Engineering and Computer Science
Department

University of Beijing, Computer Science Department

University of Nanjing, Computer Science Department

Electronics

Chinese Academy of Sciences
Chengdu Institute of Optics and Electronics
Factory 109, Beijing
Institute of Physics, Beijing
Institute of Semiconductors, Beijing
Shanghai Institute of Metallurgy
Shanghai Institute of Organic Chemistry
Xinxiang Semiconductor Device Factory (in Hebei)

Fudan University, Shanghai, Integrated Circuits Laboratory and Physics Department

Ministry of Electronics Industry
Over 50 research establishments, including First, Second, Third, . . . , Tenth Design
Institutes, and '1411' Research Institute, Beijing

Normal University of Beijing, Physics Department

Qinghua University, Beijing, Radio Electronics Department

Shanghai and Xian Polytechnic Universities, Electronic Engineering Departments

University of Beijing, Electronics Factory under Physics Department

University of Nanjing, Radio Physics Department

Railways

Ministry of Railways
Academy of Railway Sciences, Beijing: Railway Design, Railway Construction,
Desert, Track, Locomotive and Rolling Stock, Chemical Metallurgy, Steel
Construction, Signalling and Communications, Scientific and Technical Information
plus seven other research departments; Laboratories of Brake, Bridge, Concrete,
Diesel, and Soil Mechanics
Changsha College of Railways
Close affiliations with Shanghai and Xian Polytechnic Universities.

Telecommunications

Chinese Academy of Sciences
Institute of Electronics, Beijing
Institute of Physics, Beijing, Laser Division
Institute of Semiconductors, Beijing, Laser Beam Originating Devices and
Microwave Devices Divisions
Wuhan Institute of Physics, Ionosphere Physics Division
Xian Institute of Optics and Precision Mechanics

Ministry of Posts and Communications
Academy of Posts and Telecommunications Research: Research Institutes for Satellite
Communications, Beijing; Radio Wave Propagation, Xian; Microwave
Communications, Chengdu; Telephone Systems, Beijing; etc
Beijing College of Posts and Telecommunications
Chengdu Telecommunication Engineering College
Northwestern College of Telecommunications, Xian

Ministry of Radio and Broadcasting
Broadcasting Research Institute, Beijing

Nanjing College of Engineering

Qinghua University, Beijing, Radio Electronics Department

University of Beijing, Physics Department

University of Nanjing, Electronic Engineering Department

Index of Establishments

(U = University; C = College)

Subject Index

This index covers the subject matter of scientific and technological research and development. The terminology used is based on ROOT, the British Standards Institution thesaurus (1981).